International Conference on Polarised Neutrons for Condensed Matter Investigations 2014 (PNCMI 2014)

Journal of Physics: Conference Series Volume 711

Sydney, Australia
15 – 19 September 2014

Editors:

Wai Tung Hal Lee
Garry McIntyre
Kirrily Rule

ISBN: 978-1-5108-2422-5
ISSN: 1742-6588

Printed from e-media with permission by:

Curran Associates, Inc.
57 Morehouse Lane
Red Hook, NY 12571

Some format issues inherent in the e-media version may also appear in this print version.

Printed by Curran Associates, Inc. (2016)

For permission requests, please contact the Institute of Physics at the address below.

Institute of Physics
Dirac House, Temple Back
Bristol BS1 6BE UK

Phone: 44 1 17 929 7481
Fax: 44 1 17 920 0979

techtracking@iop.org

Additional copies of this publication are available from:

Curran Associates, Inc.
57 Morehouse Lane
Red Hook, NY 12571 USA
Phone: 845-758-0400
Fax: 845-758-2633
Email: curran@proceedings.com
Web: www.proceedings.com

Table of contents

Volume 711

International Conference on Polarised Neutrons for Condensed Matter Investigations (PNCMI 2014)
15–19 September 2014, Sydney, Australia

Accepted papers received: 13 April 2016
Published online: 11 May 2016

Preface

011001
OPEN ACCESS
10th International Conference on Polarised Neutrons for Condensed Matter Investigations (PNCMI)

011002
OPEN ACCESS
Peer review statement

Papers

History

012001
OPEN ACCESS
The strange history of polarised neutrons in Australia

T. J. Hicks.....1

Experiments and Applications

012002
OPEN ACCESS
Spin disorder in maghemite nanoparticles investigated using polarized neutrons and nuclear resonant scattering

M. Herlitschke, S. Disch, I. Sergueev, K. Schlage, E. Wetterskog, L. Bergström and R. P. Hermann.....12

012003
OPEN ACCESS
Structure of hydrogenous liquids: separation of coherent and incoherent cross sections using polarised neutrons

A Stunault, S Vial, L Pusztai, G J Cuello and L Temleitner.....23

012004
OPEN ACCESS
The study of perpendicular magnetic anisotropy in the magnetic sensors with linear sensitivity using polarized neutron reflectometry

T. Zhu.....28

012005
OPEN ACCESS
Enhanced Magnetism in Field-Cooled [$Ni_{80}Fe_{20}$/Mn]$_3$ Multilayers Studied Using Polarized Neutron Reflectometry

W. Uilhoorn, S. J. Callori, D.L. Cortie, H.-C. Su, Y. Khaydukov, K.-W. Lin and F. Klose.....35

012006
OPEN ACCESS
Quantum entanglement in manganese(II) hexakisimidazole nitrate: on electronic structure imaging - A polarized neutron diffraction and DFT study

Warren A. Wallace.....41

Instruments and Techniques

012007
OPEN ACCESS
Development of a ^{3}He nuclear spin flip system on an in-situ SEOP ^{3}He spin filter and demonstration for a neutron reflectometer and magnetic imaging technique

H Hayashida, T Oku, H Kira, K Sakai, K Hiroi, T Ino, T Shinohara, T Imagawa, M Ohkawara, K Ohoyama, K Kakurai, M Takeda, D Yamazaki, K Oikawa, M Harada, N Miyata, K Akutsu, M Mizusawa, J D Parker, Y Matsumoto, S Zhang, J Suzuki, K Soyama, K Aizawa and M Arai.....56

012008
OPEN ACCESS
SEOP polarized ^{3}He Neutron Spin Filters for the JCNS user program

Earl Babcock, Zahir Salhi, Tobias Theisselmann, Denis Starostin, Johann Schmeissner, Artem Feoktystov, Stefan Mattauch, Patrick Pistel, Aurel Radulescu and Alexander Ioffe.....65

012009
OPEN ACCESS
VERITAS: a high-flux neutron reflectometer with vertical sample geometry for a long pulse spallation source

S. Mattauch, A. Ioffe, D. Lott, A. Menelle, F. Ott and Z. Medic.....76

012010
OPEN ACCESS
Polarisation Analysis Neutron Spectrometer, POLANO, at J-PARC - Concept and Magnetic Field Optimisation

K. Ohoyama, T. Yokoo, S. Itoh, M. Nanbu, K. Iwasa, M. Ohkawara, N. Kaneko, T. Ino, H. Hayashida, T. Oku, H. Kira, S. Tasaki, M. Takeda, H. Kimura and T. J. Sato.....86

012011
OPEN ACCESS
Conceptual design of a polarized ^{3}He neutron spin filter for polarized neutron spectrometer POLANO at J-PARC

T Ino, K Ohoyama, T Yokoo, S Itoh, M Ohkawara, H Kira, H Hayashida, K Sakai, K Hiroi, T Oku, K Kakurai and L J Chang.....94

012012
OPEN ACCESS
Measuring gas temperature during spin-exchange optical pumping process

E Normand, C Y Jiang, D R Brown, L Robertson, L Crow and X Tong.....104

012013
OPEN ACCESS
First result from the magic-PASTIS using large ^{3}He SEOP-polarized GE180 doughnut cell

Zahir Salhi, Earl Babcock, Ramil Gainov, Klaus Bussmann, Hans Kaemmerling, Patrick Pistel, Margarita Russina and Alexander Ioffe.....110

012014
OPEN ACCESS
In-situ compact ^{3}He neutron spin polarizer based on a magneto-static cavity with built-in NMR coils

S Lee, M K Moon, J Kim, S J Cho, J H Lee, C-H Lee, S W Lee and T Ino.....119

012015
OPEN ACCESS
Magnetic field optimization and design of a superconducting neutron Wollaston prism

F. Li, S. R. Parnell, T. Wang, D. V. Baxter and R. Pynn.....127

012016
OPEN ACCESS
Interaction of neutrons with a birefringent medium moving with an acceleration

A.I. Frank.....138

PREFACE

The 10th International Conference on Polarised Neutrons for Condensed Matter Investigations (PNCMI) was organised by the Australian Nuclear Science and Technology Organisation (ANSTO) and held in Sydney, Australia on 15–19 September, 2014. Sixty-two researchers, post-docs, and students from 13 regions and countries participated in the conference. A total of 41 talks and 27 posters were presented covering the latest developments in scientific research using polarised neutrons and polarised neutron instrumentation. The topics covered included:

- Multiferroics
- Strongly correlated electron systems
- Imaging
- Soft matter and biology
- Molecular magnetism
- Magnetic nano-materials
- Thin films and multilayers
- Instrumentation development
- Facility reports

The first PNCMI was held at Dubna, Russia in 1996. It has since been held every two years in locations in Europe, Japan and the US. It was the first time for the meeting to be held in Australia. The conference's honorary chair was Dr Trevor Hicks of Monash University, who pioneered polarised neutron scattering in Australia.

We would like to acknowledge the financial support by the following research institutes: ANSTO, Oak Ridge National Laboratory in US, ISIS, Science & Technology Facilities Council in UK, Jülich Center for Neutron Science, Forschungszentrum Jülich in Germany, Institut Laue-Langevin in France, and Helmholtz Zentrum Berlin in Germany. We are grateful for the outstanding support provided by the ANSTO staff. We wish our participants and the organisers of the next PNCMIs success in their respective endeavours.

PNCMI Organising Committee
Wai Tung Hal Lee, Conference Chair, Proceedinsgs Editor
Elliot Gilbert
Shane Kennedy
Garry McIntyre, Proceedings Editor
Kirrily Rule, Proceedings Editor

International Advisory Committee

Ken Andersen	European Spallation Source
Masa Arai	Japan Atomic Energy Agency
Julie Borchers	National Institute of Standards and Technology
Stewart Campbell	University of New South Wales at Canberra
Pierre Courtois	Institut Laue-Langevin
Jason Gardner	National Synchrotron Radiation Research Center, Taiwan and ANSTO
Arsen Goukassov	Laboratoire Léon Brillouin
Sergey Grigoriev	Petersburg Nuclear Physics Institute
Klaus Habicht	Helmholtz-Zentrum Berlin
Takashi Ino	KEK Institute of Materials Structure Science
Alexander Ioffe	Jülich Centre for Neutron Science
Kazuhisa Kakurai	Quantum Beam Science Center, Japan Atomic Energy Agency
Frank Klose	Australian Nuclear Science and Technology Organisation
Sean Langridge	ISIS, STFC Rutherford Appleton Laboratory
Valeria Lauter	Oak Ridge National Laboratory
Eddy Lelièvre-Berna	Institut Laue-Langevin
Yuntao Liu	China Advance Research Reactor
Andreas Michels	University of Luxemburg
Catherine Pappas	Delft University of Technology
Roger Pynn	Oak Ridge National Laboratory and Indiana University
Virginie Simonet	Institut Néel
Tony Xin Tong	Oak Ridge National Laboratory

Past PNCMI meetings

1996 Dubna
1998 Grenoble
2000 Gatchina
2002 Jülich
2004 Washington DC
2006 Berlin
2008 Tokai
2010 Delft
2012 Paris

Peer review statement

All papers published in this volume of *Journal of Physics: Conference Series* have been peer reviewed through processes administered by the proceedings Editors. Reviews were conducted by expert referees to the professional and scientific standards expected of a proceedings journal published by IOP Publishing.

The strange history of polarised neutrons in Australia

T. J. Hicks

School of Physics, PO box 27, Monash University, Victoria 3800, Australia

trevor.hicks@monash.edu

Abstract. The history of polarised neutrons in Australia is unusual firstly because of the particular access that individuals in universities had to the facilities at the reactor site and because this resulted in the experiments being done almost all being with polarisation analysis. Two instruments were initially available. One was a conventional instrument albeit with a tilting counter. The other was a primitive polarisation analysis instrument purpose built for diffuse scattering. This latter instrument evolved over more than thirty years and produced results ranging from the separation of magnetic and nuclear diffuse scattering, for which it was conceived, to the isolation of magnetic features in inelastic spectra.

1. Introduction

The first polarised neutron experiments at the HIFAR reactor in Sydney were unusual in several respects. The first was in the method of accessing the facilities on the reactor site and the second was that the experiment was a polarisation analysis one.

In 1958 the Australian Institute of Nuclear Science and Engineering (AINSE) was formed. This was to be a bridge between the Australian Atomic Energy Commission (AAEC) and the universities. The then chairman of the AAEC set out the philosophy at the inauguration of AINSE. He said

"Not even the wealthiest university can equip itself completely for the study of nuclear energy today. The universities provide basic training in nuclear science and engineering with their own equipment and staff, but in any advanced programme they must have access to a nuclear reactor and specialised chemical, engineering and metallurgical laboratories to handle highly radioactive materials." J P Baxter.

The role of AINSE was: To provide the universities access to the facilities of the AAEC and to provide staff and students of those travel and accommodation to access those facilities. The AAEC and the universities provided funds and this resulted in Australia having the world's first user programme for neutron beams among other facilities. Proposals were sought and peer reviewed and accepted or rejected.

It should be emphasised that this user programme was for individuals within universities and not for the universities themselves. The result was that the university or AAEC bureaucracies had no influence on the success or failure of particular proposals.

In the following sections we will outline how it was that the first polarised neutron experiment was a polarisation analysis experiment rather than a simple polarised beam one.

2. First polarised neutron instrument

By the late 1960s more instruments and support were required to satisfy demand than could be provided by the AAEC. AINSE then formed its own neutron diffraction group consisting of Frank

Moore, David Wheeler and Roy Ebdon. One of their first projects was to build a versatile polarized neutron diffractometer.

At that time there was a potential demand for such an instrument from Brian Figgis, a computational chemist from the University of Western Australia. So the instrument was built by the AINSE group. It was a classic instrument with a CoFe monochromator polarizer. It had a vertical field at the specimen position and a tilting counter to access other reflection zones. Figure 1 shows the instrument in 1972 [1]

Firure 1. The first polarised neutron diffractometer at Lucas Heights. Note tilting detector. The photo was taken in 1972.

As it happened Brian Figgis never used this instrument. Instead he collaborated with colleages from ILL.

3. LONGPOL I

At almost the same time another polarised neutron instrument was being constructed. This resulted from a proposal to AINSE to isolate the magnetic diffuse scattering in the dilute alloy systems Cu-Mn and Cu_Fe. Cu-Mn was widely regarded as a spin glass and it was known that that there was cusp in the magnetic susceptibility but it was not at all clear whether this was due to some degree of magnetic ordering or a simple freezing.

The new instrument was based on a previously existing helical velocity selector with out of pile shielding and a wavelength resolution of 10%. The diffractometer itself was single axis and the detector was heavily shielded and rested on airpads on the floor. The detector was rotated by manually pushing it while the airpads were pressurised.. Polarisation and analysis were achieved by polycrystalline iron filters in electromagnets before and after the specimen. It was important that the filters were magnetically saturated. The result was that the Bragg scattering in the filterswas different for up and down neutrons which discrimination peaking at the Bragg cutoff. Immediately below the Bragg cutoff (3.6 Å) the cross sections for up and down spin neutrons were in a ratio 2:1 which for comparison is equivalent to a He^3 filter with a 65% He^3 nuclear polarisation. As for He^3 polarisers the thickness of the filters is a compromise between perfect transmission and perfect polarisation. The iron filters have broad angular acceptance and relatively broad wavelength acceptance which is very good for a diffuse scattering instrument, like LONGPOL, not requiring good resolution. Stewart Campbell, Naeem Ahmed and I surveyed many ferromagnetic and ferrimagnetic materials by calculating their up and down neutron spin cross sections but we found none better than iron [2,3]. Finally LONGPOL had an RF flipper mounted before the specimen so that the incident beam could be spin flipped. To separate magnetic from nuclear scattering the polarisation was directed along the scattering vector at the specimen.

Figure 2 shows the instrument with Naeem Ahmed behind

Figure 2. The first LONGPOL in 1972. The detector was positioned manually on the airpads.

LONGPOL I provided the first direct evidence that Cu-Mn was a spin glass. A cusp in the variation with temperature had been observed in the magnetic susceptibility. However this was similar to that observed for the antiferroagnetic critical temperature. No long range order had been observed and it had been surmised that the magnetic distribution was glass like. LONGPOL was able to directly measure the magnetic diffuse scattering [4] and separate it from the scattering from the nuclei and the results are shown in figure 3.

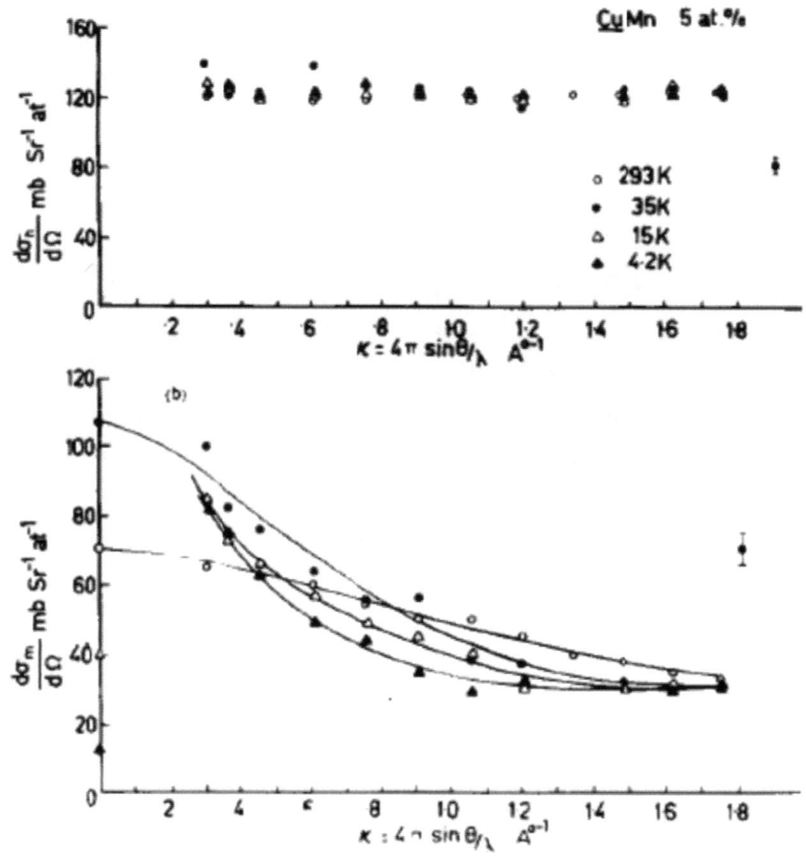

Figure 3. Diffuse scattering from Cu 5 at% Mn. (a) Nuclear cross section. (b) Magnetic scattering at various temperatures. Susceptibility values are also shown for the various temperatures. A typical error bar is shown.

Plot (a) shows the nuclear scattering and plot (b) shows the magnetic scattering. The nuclear scattering is flat as function of scattering vector indicating that there is a random distribution of Cu and Mn nuclei, but there is some correlation in the magnetic scattering which develops slightly with lower temperature.

Furthermore the small scattering vector cross section of the magnetic scattering was compared with the bulk susceptibility. The cross sections were placed on an absolute basis and the cross sections derived from the susceptibility plotted at zero scattering vector for each of the scans taken. It can be seen that at the two lower temperatures any extrapolation to zero scattering vector of the scattering will not match the susceptibility measurements. At the two higher temperatures extrapolation of the scattering does match the susceptibility. This means that at high temperatures the magnetic fluctuations are free to respond to a magnetic field. At lower temperatures a large fraction of the magnetic fluctuations are too slow to respond to a magnetic field, that is, they are frozen or glass like. Further Cu-Mn experiments and experiments on similar alloy systems were also performed at this time.

4. LONGPOL II

LONGPOL was shifted to a more intense beam and the wavelength selection was changed from a velocity selector to a graphite double monochromator. Apart from this the basic diffractometer remained the same. Figure 4 shows the modified instrument.

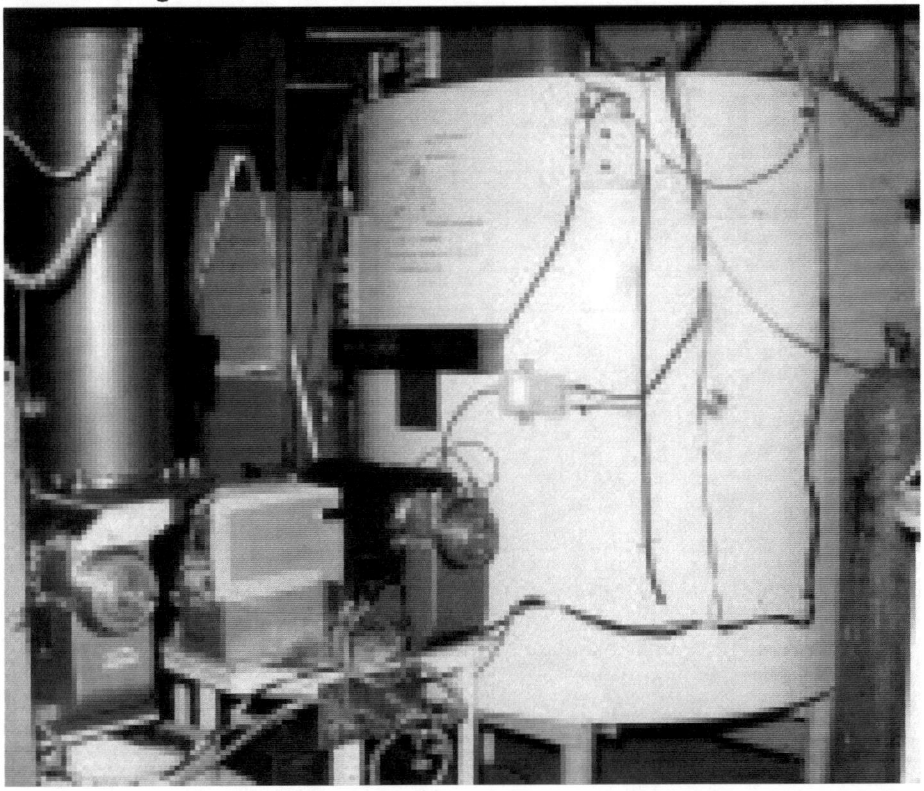

Figure 4. LONGPOL II. The change from LONGPOL I is a more intense beam hole and double graphite monchomator. The out of pile shielding retrieved from the junkyard is shown.

The out of pile shielding was retrieved from the AAEC's junkyard. The RF flipper and cryostat are shown more clearly in this photo.

The other major modification was to select elastic scattering. To achieve this the RF flipper was pulsed randomly and the detector was gated with the same pulse sequence delayed at the neutrom time of flight. This was implemented by John Davis. Just to confirm that it was effective, a magnon in antiferromagnetic Mn-Cu was measured in neutron energy gain by altering the delay [5].

The investigation of magnetic diffuse scattering was extended to antiferromagnets at this time. An example is that from a single crystal of antiferromagnetic Mn-Ni alloy as seen in figure 5 The left panel shows the separated nuclear scattering from the crystal along these directions. Clearly the distribution of atoms is not random but the distribution statistics can be obtained from these measurements. The right panel shows the magnetic diffuse scattering which reflects both the distribution of magnetic atoms and the variation of magnetic moment due to their environment [6]. The striking feature is the peak in diffuse scattering which occurs near the 001 position. Assuming the antiferromagnetic structure is collinear the absence of a Bragg peak at 001 indicates that the spin direction of the structure is along z. The magnetic fluctuations at this point must therefore be transverse in order to contribute to the diffuse scattering.

Journal of Physics: Conference Series **711** (2016) 012001 doi:10.1088/1742-6596/711/1/012001

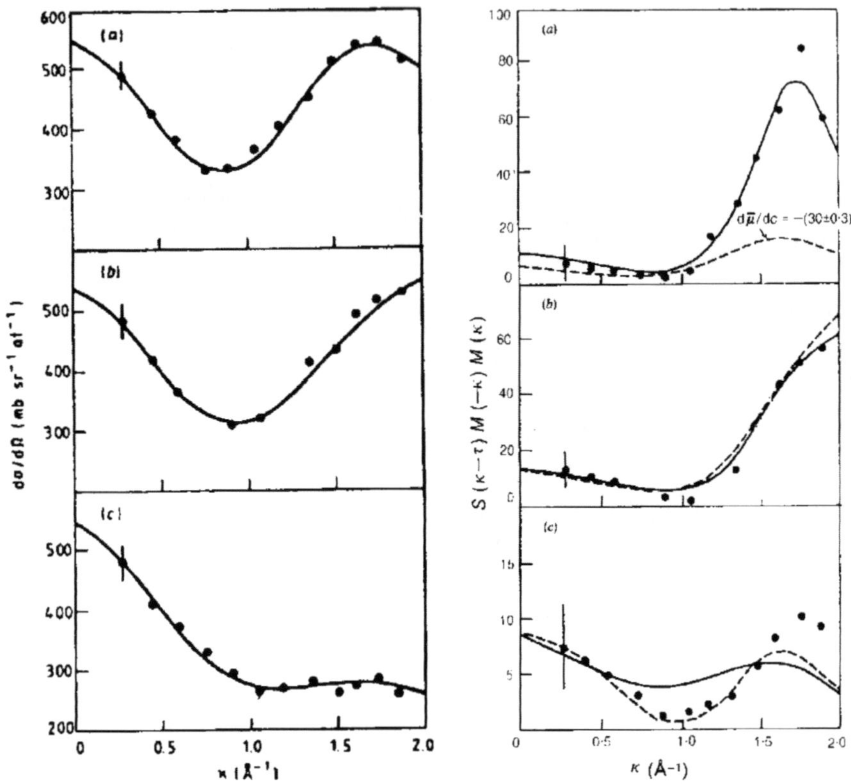

Figure 4. Diffuse scattering from antiferromagnetic single crystal Mn Ni. On the left is the nuclear scattering. The magnetic scattering is on the right. (a) 001 direction. (b) 110 (c) 111.

LONGPOL III

The makeover of LONGPOL at this point was an attempt, after fifteen years, to engineer the instrument. A further move to a yet more intense beam was made however with a more than ideal length of throw between monochromators due to the constraints of the reactor beam hall. Figure 5 shows the layout of the new instrument.

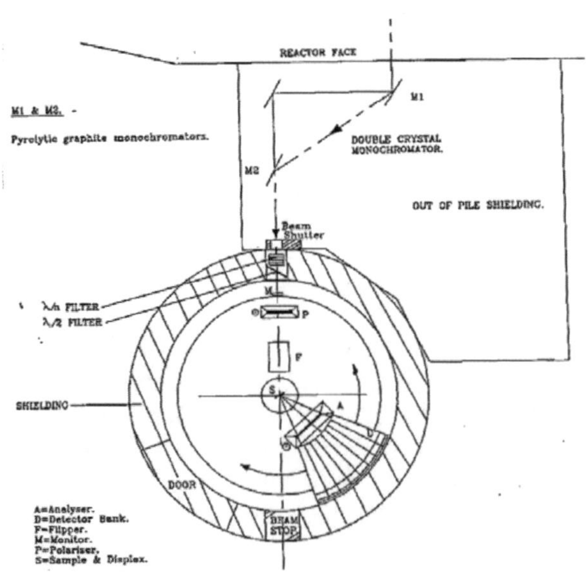

Figure 5. Layout of LONGPOL III

From the double graphite monochomator the beam passes from the out of pile shielding into the instrument's barrel shielding and through the higher order filters. The beam is the polarized by an iron filter and flipped by a Mezei coil before passing through the specimen position with the ability to turn the polarisation horizontal. The scattered beam is analysed by a magnetized iron strip before being detected by one of the eight detectors. Barry Maguire developed the software control of the instrument. The polarizing and analyzing filters would later be replaced by supermirrors. The building of the instrument was a team effort. The AAEC (Ansto) provided the out of pile shielding and the diffractometer table with the detector arm. AINSE provided the barrel shielding and Monash University the analyser magnet with the Helmholtz coils which provided the background field and which was financed by national competitive grants. The instrument with the iron filters is shown in figure 6 with Shane Kennedy looking on.

Figure 6. LONGPOL III with Shane Kennedy. The edge of the electromagnet holding the analyzing iron filter can just be seen.

At this point it is appropriate to tell the story of the implementation of supermirror polarisers. A supermirror polarizer was first fabricated in 1982. Shane Kennedy while a research assistant at Monash University used an old evaporator to deposit layers of Fe-Co and Al on a silicon wafer which provided good reflectivity and polarization out to $2\theta_c$. Fifteen years later we were able to arrange to have similar supermirrors sputtered at HMI for incorporation into stacks for the detectors and for a polarizer. A stack is shown in figure 7

Figure 7. Compact supermirror assembly. The stack is held in the red permanent magnet. In the front is the cam to adjust the silicon stack curvature.

The permanent magnet provided the field to magnetise the layers and a cylindrical cam provided the the ability to adjust the curvature of the wafers. Stack dimensions were 7x5x2 cm. Figure 8 summarises the stack performance.

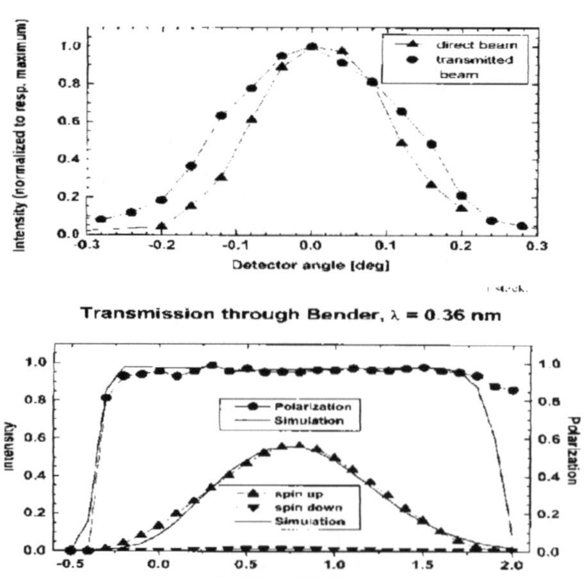

Figure 8. Performance of a supermirror bender stack.

Journal of Physics: Conference Series **711** (2016) 012001 doi:10.1088/1742-6596/711/1/012001

Basically the stacks had 57% transmission of the right spin, close to 0% transmission of the wrong spin with 97% polarization over $2\theta_c$ with an acceptance angle of 1° at 3.6 Å [7].

There were many diffuse scattering experiments performed with this instrument but some results highlighting the wide coverage of phase space in the investigation of spin orientation in a mixed ferromagnetic antiferromagnetic structure. Figures show the peaks in ferromagnetic Fe_2MnSi from the antiferromagnetic modulation. Lack of a 111 means that the antiferromagnetic direction is <111> and because the <113> is completely spin flip the antiferromagnetic modulation is collinear [8].

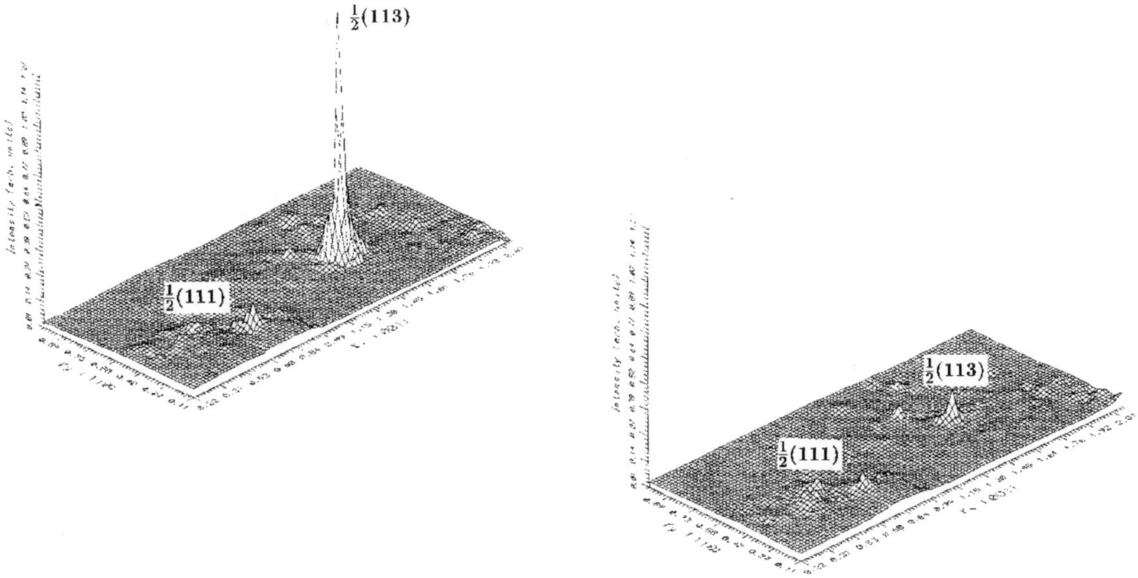

Figure 9. Bragg scattering fom antiferromagnetically modulated ferromagnetic Fe_2MnSi. Spin flip scattering on the left. Non spin flip scattering on the right.

With increased intensity and multidetectors some time of flight experiments were attempted with pseudo random pulsing of the flipper. Figure 11 shows a spectrum of crystal field excitations in PrAl3 taken with the iron filters still in place. With the neutron spin along the scattering vector all magnetic features are with spin flip whereas the features with no spin flip are from nuclear scattering. Consequently the magnetic features in the spectrum are negative going and the nuclear features are positive going.

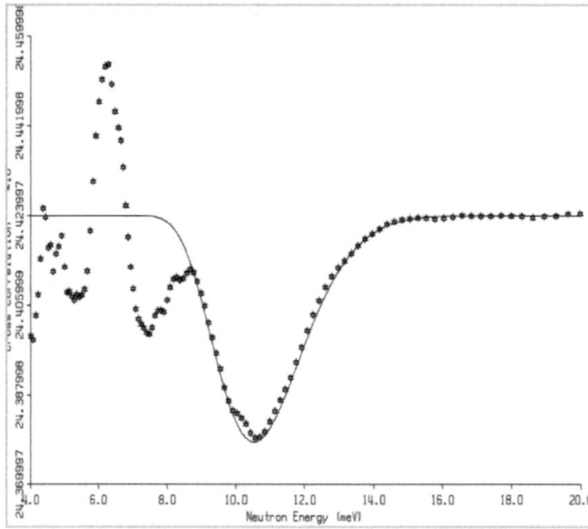

Figure 10. Spectrum of $PrAl_3$. Positive going features are magnetic. Negative going features are nuclear

The spectrum shows the nuclear elastic peak at 6.0 meV against a magnetic quasielastic peak. At a neutron energy of 10.5 meV there are two barely resolved crystal field excitations [9].

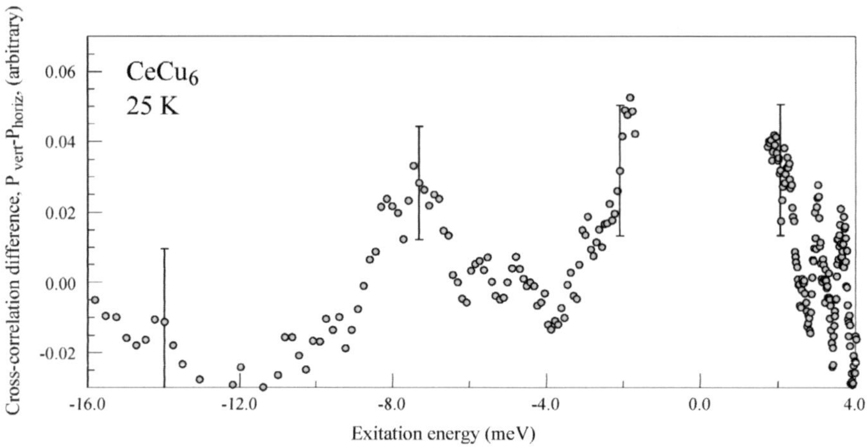

Figure 11. Magnetic inelastic scattering from $CeCu_6$ at 25 K.

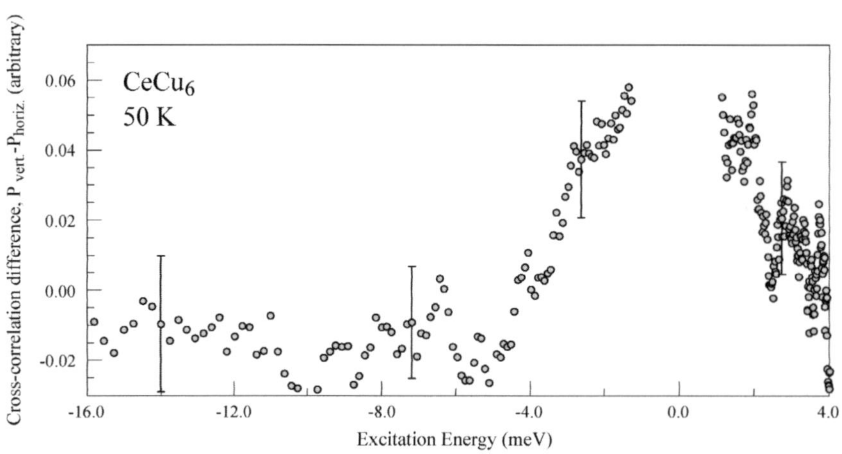

Figure 12. Magnetic inelastic scattering from $CeCu_6$ at 50 K.

Later a more serious attempt was made to separate out the crystal field spectrum from $CeCu_6$. Spectra were taken with the neutron spin along the scattering vector and at right angles perpendicular to the scattering plane and the difference taken. This should, apart from nuclear spin scattering, be only magnetic scattering. The error bars were obtained by collecting many data sets for each temperature and for each neutron spin orientation, cross correlating each, averaging and taking the standard deviation at each point.

The spectra show quasielastic scattering at each temperature but broader at 50 K. There is also an excitation at 7 meV at 25 K which is rather washed out at 50 K. These are consistent with other unpolarised studies but now these features are clearly identified as magnetic [10].

Conclusion and present day

From 1971-2, when the two instruments incorporating polarised neutrons were first built at the HIFAR reactor there has been a continuous polarised neutron programme at Lucas Heights only interrupted by final shut down of HIFAR. With the advent of the OPAL reactor there has been an intention to provide polarised neutrons on a wide range of instruments. At time of writing there are polarised neutrons on several instruments.

There is currently a programme to provide He^3 filters for wide use but the most ambitious project is to provide analysis covering the whole detector bank of the cold neutron time of flight instrument PELICAN (son of LONGPOL). When this is complete PELICAN will be able to do many of the experiments attempted on LONGPOL but with orders of magnitude more intensity.

References

[1] F H Moore, *Atomic Energy* April 1972, 2.
[2] S.J.Campbell, N.Ahmed, T.J.Hicks, R.Ebdon and D.A.Wheeler `LONGPOL I: *J.Phys.* E7, 195 (1974).
[3] N.Ahmed, S.J.Campbell and T.J.Hicks `LONGPOL II: *J.Phys.* E7, 199 (1974).
[4] N.Ahmed and T.J.Hicks *Solid State Comm.* **15**, 415 (1974).
[5] J.R.Davis, R.Cywinski, O.Moze and T.J.Hicks *J.de Phys.*C7, 65 (1982).
[6] O Moze, T J Hicks (1981), *J Phys* F **12**, 1
[7] Th Krist, S J Kennedy, T J Hicks, F Mezei (1998), *Physica* B **241-3**, 82
[8] T Ersez, S J Kennedy, T J Hicks, *J Phys CM* **7**, 8423
[9] D J Goossens, S J Kennedy, T J Hicks (1996), *Nuc Inst Meth* A **380**, 572
[10] S. J.Harker, T. J. Hicks, S. J. Kennedy, Dehong Yu, *Physica B* **397**, 15 (2007).

Spin disorder in maghemite nanoparticles investigated using polarized neutrons and nuclear resonant scattering

M. Herlitschke[1,2,3], S. Disch[4,5], I. Sergueev[1], K. Schlage[1], E. Wetterskog[6], L. Bergström[7], R. P. Hermann[2,3,8]

[1] FS-PE, Deutsches Elektronen-Synchrotron, 22607 Hamburg (Germany)
[2] JCNS and PGI, JARA-FIT, Forschungszentrum Jülich GmbH, 52425 Jülich (Germany)
[3] Faculté des Sciences, Université de Liège, 4000 Liège (Belgium)
[4] Department Chemie, Universität zu Köln, 50939 Köln (Germany)
[5] Institut Laue-Langevin, 38042 Grenoble (France)
[6] Department of Engineering Sciences, Angström Laboratory, 75121 Uppsala (Sweden)
[7] Department of Materials and Environmental Chemistry, Stockholm University, 10691 Stockholm (Sweden)
[8] Materials Science and Technology Division, Oak Ridge National Laboratory, 37831 Oak Ridge, Tennessee (USA)

E-mail: marcus.herlitschke@desy.de, sabrina.disch@uni-koeln.de

Abstract. The manuscript reports the investigation of spin disorder in maghemite nanoparticles of different shape by a combination of polarized small-angle neutron scattering (SANSPOL) and nuclear forward scattering (NFS) techniques. Both methods are sensitive to magnetization on the nanoscale. SANSPOL allows for investigation of the particle morphology and spatial magnetization distribution and NFS extends this nanoscale information to the atomic scale, namely the orientation of the hyperfine field experienced by the iron nuclei. The studied nanospheres and nanocubes with diameters of 7.4 nm and 10.6 nm, respectively, exhibit a significant spin disorder. This effect leads to a reduction of the magnetization to 44 % and 58 % of the theoretical maghemite bulk value, observed consistently by both techniques.

1. Introduction

Magnetic nanomaterials are nowadays used in a wide range of applications, *e.g.*, spintronics, cell labelling for magnetic separation, magnetic resonance imaging contrast enhancement, drug delivery or magnetic hyperthermia[1, 2, 3]. A large number of these applications rely on magnetic interparticle interactions or on intraparticle magnetism, motivating further investigation of the intraparticle spin structure. The overall magnetic moment of small particles is typically lower than the bulk value[4, 5], a phenomenon largely attributed to surface effects[6, 7, 8] with a surface layer often thought of as a spin glass[9, 10, 11]. Whereas a core-shell model of spin canting is often used for simplicity and conveniently reducing the number of free parameters in modelling[12, 13], recent studies of the spatial magnetization distribution in magnetic nanoparticles suggest that spin disorder in the nanoparticle core plays a major role for the magnetization of the particle[14].

Here, we present studies of reduced magnetization and the concomitant spin disorder in iron oxide nanoparticles of different shape by application of two different techniques: Polarized

Small-Angle Neutron Scattering (SANSPOL) and Nuclear Forward Scattering (NFS). These methods give access to different information, as SANSPOL is sensitive to the spatially resolved magnetization distribution and NFS is indirectly sensitive to the spin orientation of iron atoms influencing the nuclear hyperfine field in the nanomaterial. Our results indicate a significant degree of spin disorder in the studied nanoparticles, observed with both techniques, in agreement with macroscopic measurements and the literature.

2. Experimental

Particle Synthesis

Maghemite nanoparticles of spherical and cubic shape were prepared according to a thermolytic, oleic-acid-assisted synthesis route described previously [15, 16, 17]. For preparation of ^{57}Fe-enriched nanoparticles, the starting material $^{57}FeCl_3$ was prepared by heating elemental ^{57}Fe in an anhydrous Cl_2 gas stream[18]. Anhydrous Cl_2 gas was produced by oxidation of concentrated hydrochloric acid with $KMnO_4$. The evolving Cl_2 gas stream was washed from excess HCl using H_2O, dried over concentrated H_2SO_4, and subsequently passed over pieces of ^{57}Fe (0.5 g, 8.8 mmol) heated to 250°C in inert atmosphere. $^{57}FeCl_3$ (1.208 g, 7.4 mmol, 84 %) was collected as a dry, but strongly hygroscopic, shiny violet powder which turned quickly into a yellow paste when subjected to air. Mössbauer spectroscopy confirmed the high purity of the produced nanoparticles, see the supplementary data.

Morphological Characterization

Transmission electron microscopy (TEM) images were obtained using a JEOL JEM-2100 microscope equipped with a LaB_6 filament operated at 200 kV (f = 2.7 mm, Cs = 1.4 mm, Cc = 1.8 mm, and point resolution 2.5 Å). Small-angle X-ray Scattering (SAXS) was carried out at the B1 beamline of HASYLAB/DESY using an incident X-ray energy of 8.92 keV. The small-angle scattering was recorded using a Pilatus 300K detector set up at 0.935 and 3.635 m distances from the sample position. The data measured by dilute nanoparticle dispersions in toluene (7 mg/ml) were averaged circularly and analyzed according to spherical particle form factors.

SANSPOL

Polarized Small-angle Neutron Scattering (SANSPOL) data was collected at the D33 instrument at the Institut Laue-Langevin, using 6 Å neutron wavelength and 5.3 m detector and collimation distance. Dilute dispersions (7 mg/ml) of nanospheres and nanocubes in toluene (deuterated by 80 %) were measured with a polarized incident neutron beam ($P = 0.826$) in a horizontally applied magnetic field of 1 T. Data correction to the scattering background of the pure solvent and normalization to absolute intensities were carried out using the SANS data reduction software GRASansP[19]. The purely nuclear (I_N) as well as the nuclear-magnetic (I(+) and I(-)) scattering cross sections were radially integrated in 10° sectors oriented parallel and perpendicular to the applied magnetic field, respectively.

NFS

Nuclear Forward Scattering (NFS) measurements were performed at beam line P01 of the large scale facility PETRA III. The nanoparticles were dispersed in toluene with a concentration of $\approx$ 14 mg/ml and put into a sample holder with 8 mm thickness. The sample was inserted in the pre-cooled cryostat, in order to obtain a rapid freezing of toluene, with the field-cooled (FC) procedure and a magnetic field of 0.7 T, reaching a sample temperature of 15 – 30 K. The FC procedure reproduces similar conditions as for the SANSPOL measurements, because the particle easy axis will align parallel to the field at room temperature and the solution will freeze in this state on passing the solidification point of toluene. The magnetic field was sequentially applied, through rotation of the magnet, along the x, y and z direction.

3. Theory

Nuclear Forward Scattering - Fitting Model

The spin canting can be studied in Mössbauer spectroscopy by analyzing the ratio of different lines in a spectrum, which gives an average direction of the local magnetic field. Here, we introduce a similar approach to investigate the nuclear-forward-scattering time spectra, where an average spin direction can be obtained without a preliminary introduced model. The amplitude of the scattering consists of several components defined by the principal axis system of the polarized synchrotron beam. The ratio of these components describes the relative orientation of the hyperfine field to these axes. As shown in this section, the projections on x, y and z are useful and enable to continuously interpolate between spins randomly oriented in 3D and spins fully aligned in a specific direction.

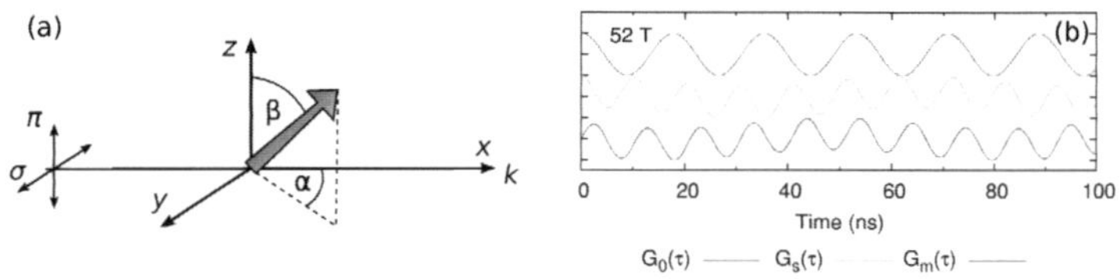

Figure 1. (a) Sketch of the coordinate system. The red arrow indicates the direction of an arbitrary hyperfine field, determined by the angles α and β. The propagation vector of the beam, k, is along the x direction, σ and π polarizations are along y and z, respectively. (b) Schematic representation of the G-terms, described in the text, for an iron nucleus in a hyperfine field of 52 T.

The NFS *amplitudes* for linear-polarised incoming radiation in case of a static hyperfine magnetic field and thin samples can be derived from Ref. [20, 21]. These amplitudes, for the chosen coordinate system depicted in figure 1, are:

$$A_{\sigma\sigma} = \frac{1}{2}(G_0(\tau) + G_s(\tau) + [G_0(\tau) - G_s(\tau)]\cos 2\beta) \tag{1}$$

$$A_{\sigma\pi} = G_m(\tau)\cos\alpha\sin\beta - (G_0(\tau) - G_s(\tau))\frac{1}{2}\sin 2\beta\sin\alpha \tag{2}$$

$$G_0(\tau) = \cos\frac{\tau(\omega_e - \omega_g)}{2} \tag{3}$$

$$G_s(\tau) = \frac{3}{4}\cos\frac{\tau(3\omega_e - \omega_g)}{2} + \frac{1}{4}\cos\frac{\tau(\omega_e + \omega_g)}{2} \tag{4}$$

$$G_m(\tau) = \frac{3}{4}\sin\frac{\tau(3\omega_e - \omega_g)}{2} + \frac{1}{4}\sin\frac{\tau(\omega_e + \omega_g)}{2}, \tag{5}$$

where $A_{\sigma\sigma}$ is the amplitude without polarization change and $A_{\sigma\pi}$ is the amplitude with polarization change. The constants $\omega_e = B \cdot 4.96\,(\mu\text{s}\cdot\text{T})^{-1}$ and $\omega_g = -B \cdot 8.68\,(\mu\text{s}\cdot\text{T})^{-1}$ are the Larmor frequencies in the field B/μ_0 of the ^{57}Fe 14.4 keV excited and ground state, respectively. The perturbation factors $G(\tau)$, depicted in figure 1, describe the nuclear transitions. $G_0(\tau)$ corresponds to transitions with magnetic quantum number $M = 0$, and the $G_s(\tau)$ and $G_m(\tau)$ correspond to $M = \pm 1$. Transformed into the frequency domain, G_0 reproduces the lines 2 and 5 of a Mössbauer spectrum, and G_s and G_m reproduce the lines 1, 3, 4 and 6.

The given scattering amplitudes enable the calculation of a time spectrum for any configuration of the hyperfine field, by going from the amplitudes to the intensities with an

Table 1. Selected hyperfine field orientations and the resulting scattering amplitudes. The x^+ indicates nuclei with hyperfine field in the beam propagation direction, $x^\pm$ indicates an equal number of nuclei with hyperfine fields parallel and antiparallel to the beam propagation direction.

	$A_{\sigma\sigma}$	$A_{\sigma\pi}$
$H_{hf} \parallel z$	G_0	0
$H_{hf} \parallel y$	G_s	0
$H_{hf} \parallel x^+$	G_s	G_m
$H_{hf} \parallel x^\pm$	G_s	0
3D powder	$\frac{1}{3}(G_0 + 2G_s)$	0

additional exponential decay. Several orientations and the resulting amplitudes are compared in table 1. For example, assuming a hyperfine field along z ($\beta = 0$ and $\alpha =$ arbitrary), leads to $A_{\sigma\sigma} = G_0$ and $A_{\sigma\pi} = 0$. If several directions of the hyperfine field exist, the total amplitude will be the sum of partial amplitudes.

When this description of NFS is applied to the scattering signal of the nanoparticles, different aspects have to be considered: the spins in the maghemite particles order ferrimagnetically within one particle, forming a macro moment. But on the atomic length scale, spins are not perfectly in the ferrimagnetic order but rather canted [5, 14, 11]. The corresponding hyperfine field experienced by the ^{57}Fe nucleus is anti-parallel to the spin due to the Fermi contact term. An applied magnetic field parallel to the z axis, see figure 1, leads to orientation of the particle moment and the canted spins form a distribution that is symmetric around the applied field direction. This symmetry around the field is not a unique behaviour of the canting effect but rather a general observation, because the only present preferential direction in the sample is the field direction. Thus, the orthogonal plane behaves powder-like and all directions are energetically equal.

In order to describe this distribution of spins, a function $f(\beta)$ is assumed with $\int_0^\pi f(\beta)\sin\beta \, d\beta = 1$, expressing the hyperfine field orientation density in a particular direction, with respect to the z axis. Introducing this function $f(\beta)$ in the amplitudes $A_{\sigma\sigma}$ and $A_{\sigma\pi}$ and performing a volume integration leads to the averaged amplitudes $A^{(z)}_{\sigma\sigma}$ and $A^{(z)}_{\sigma\pi}$. The resulting G_0 and G_s terms can be separated:

$$\begin{aligned} A^{(z)}_{\sigma\sigma} =& \frac{1}{4\pi}\int_0^{2\pi} d\alpha \int_0^{\pi} d\beta \quad f(\beta)\sin\beta \; A_{\sigma\sigma} \\ =& G_0 \int_0^{\pi} d\beta \quad f(\beta)\sin\beta \; \frac{1}{4}(1+\cos 2\beta) \\ &+ G_s \int_0^{\pi} d\beta \quad f(\beta)\sin\beta \; \frac{1}{4}(1-\cos 2\beta) \\ =& G_0 a^{(z)} + G_s b^{(z)} \propto G_0 + G_s c^{(z)}_{\sigma\sigma}. \end{aligned} \tag{6}$$

The obtained parameters $a^{(z)}$ and $b^{(z)}$ are constant, since the integration over β can be performed for a given $f(\beta)$ leading to a scalar. The constant $c^{(z)}_{\sigma\sigma}$ gives a quantitative description of the function $f(\beta)$, but more than one function $f(\beta)$ can lead to the same $c^{(z)}_{\sigma\sigma}$ and thus to the same scattering signal. The same procedure can be performed for the amplitude $A_{\sigma\pi}$ leading to $A^{(z)}_{\sigma\pi} = 0$, due to vanishing integration over α. Comparing these amplitudes with table 1, one can conclude that $c^{(z)}_{\sigma\sigma} = 2$ describes the 3D random case and $c^{(z)}_{\sigma\sigma} = 0$ describes the perfect alignment parallel to z. More generally, a sum of configurations from table 1 can be used in order to describe the scattering for an arbitrary function $f(\beta)$, since only the G_0 and G_s terms are necessary in order to represent $f(\beta)$.

The same procedure can be applied for other quantization axes, *e.g.*, in case of an applied magnetic field along y or x. For sake of simplicity, the coordinate system should be changed in such way that the polar angle θ is always between the hyperfine field and the externally applied magnetic field. Consequently, the amplitudes need to be transformed to the new coordinate system. The integration over α can always be performed due to the rotation symmetry around the applied magnetic field and a description of the function $f(\beta)$ will be obtained for the new coordinate system. For the case of a magnetic field along y one obtains:

$$A_{\sigma\sigma}^{(y)} \propto c_{\sigma\sigma}^{(y)} \frac{1}{2}(G_0 + G_s) + G_s \tag{7}$$

$$A_{\sigma\pi}^{(y)} = 0. \tag{8}$$

This case is different from the former in that a complete alignment of the hyperfine field along the y axis results in $A_{\sigma\sigma}^{(y)} \propto G_s$, whereas the 3D random alignment is again $A_{\sigma\sigma}^{(y)} \propto G_0 + 2G_s$. The $\frac{1}{2}(G_0 + G_s)$ term represents a 2D alignment orthogonal to the field.

The case of a magnetic field parallel to the beam and thus along x is the most complex one, because the scattering process is sensitive to the relative amount of hyperfine fields parallel and anti-parallel to the applied field. The amplitudes are:

$$A_{\sigma\sigma}^{(x)} \propto c_{\sigma\sigma}^{(x)} \frac{1}{2}(G_0 + G_s) + G_s \tag{9}$$

$$A_{\sigma\pi}^{(x)} \propto c_{\sigma\pi}^{(x)} G_m. \tag{10}$$

A complete alignment of all hyperfine fields either parallel or anti-parallel to the beam correspond to $c_{\sigma\sigma}^{(x)} = 0$ and $c_{\sigma\pi}^{(x)} = 1$. In case the observed hyperfine fields are parallel and anti-parallel to the beam with equal amount, the constants will be $c_{\sigma\sigma}^{(x)} = 0$ and $c_{\sigma\pi}^{(x)} = 0$. The 3D random case is represented with $c_{\sigma\sigma}^{(x)} = 2$ and $c_{\sigma\pi}^{(x)} = 0$. As a remark, the amplitudes can be scaled by a constant factor in the analysis, as done before, but both amplitudes have to be scaled in the same way in order to not influence the resulting fit with the scaling.

Summarizing, the constants $c_{\sigma\sigma}^{(x,y,z)}$ were always defined such that a value of zero corresponds to a complete alignment of all hyperfine fields and thus of all spins. A value of $c_{\sigma\sigma}^{(x,y,z)} = 2$ correspond always to the 3D random case for all three field directions. In contrast, $c_{\sigma\pi}^{(x)}$ is defined such that a value of zero corresponds to an equal number of the hyperfine fields parallel and anti-parallel to the beam and a value of one denotes the case of either complete parallel or anti-parallel alignment.

4. Results and Discussion

^{57}Fe-enriched nanoparticles

Maghemite (γ-Fe_2O_3) nanoparticles of spherical and cubic shape, enriched with ^{57}Fe, were prepared according to previously published synthesis routes [15, 16, 17]. Small-angle X-ray Scattering (not shown) confirms the narrow particle size distribution inferred by TEM analysis, yielding a spherical particle radius of 3.68(1) nm with a lognormal size distribution of $\sigma_{log} = 0.057(5)$. For the nanocubes, approximated with a spherical particle form factor due to their orientation distribution in dispersion, a particle radius of 5.28(1) nm is obtained, along with a slightly larger size distribution of $\sigma_{log} = 0.070(1)$ resulting from the nonspherical shape of the particles.

The purely nuclear SANS scattering cross section (I_N) was refined according to a core-shell particle form factor, reflecting the iron oxide particle core and an oleic acid ligand shell covering

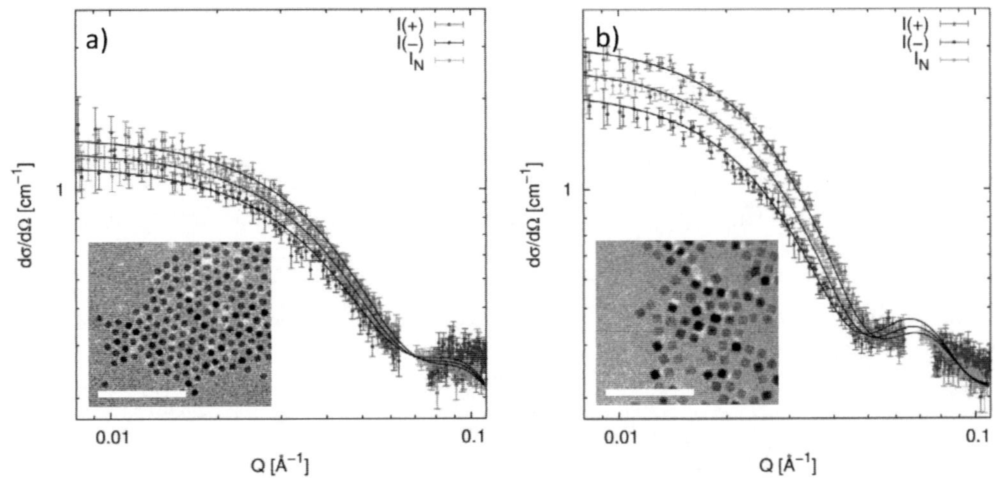

Figure 2. SANSPOL by iron oxide a) nanospheres and b) nanocubes. Purely nuclear (I_N) and nuclear-magnetic ($I(+)$, $I(-)$) cross sections, obtained from radial integration in 10° sectors oriented parallel and perpendicular to the applied magnetic field, are presented including refinements of the core-shell nanoparticle morphology. Missing data points are due to the gap between the two detectors at D33 (ILL). Insets: respective TEM images (scale bars: 100 nm)

the particle surface. The ligand shell thickness was determined as 1.5(1) nm and 2.0(1) nm for the spheres and cubes, respectively, in reasonable agreement with the maximum oleic acid ligand shell thickness of up to 2.1 nm. Moreover, quantitative analysis of the SANS cross sections shown in figure 2 additionally enables determination of the γ-$^{57}Fe_2O_3$ scattering length density of $4.96(5) \cdot 10^{-6}$ Å^{-2} which, consistently for both samples, indicates a degree of ^{57}Fe enrichment of 65(5) %.

Polarized Small-Angle Neutron Scattering

Polarized SANS (SANSPOL) was carried out in order to determine the magnetic morphology of the samples via the magnetic form factor. For a dilute system of non-interacting particles aligned in an external field, the nuclear and magnetic scattering amplitudes F_N and F_M interfere according to

$$I(\pm)_{Q,\alpha} = F_N^2(Q) + [F_M^2(Q) \mp 2\varrho^{\pm} F_N(Q) F_M(Q)] \sin^2\alpha \quad (11)$$

$$I(+)_{Q,\alpha} - I(-)_{Q,\alpha} = -2(\varrho^+ + \varrho^-) F_N(Q) F_M(Q) \sin^2\alpha \quad (12)$$

where α is the azimuthal angle between the scattering vector Q and the applied magnetic field direction, and $\varrho^+ = P, \varrho^- = P\epsilon$ with $P = 0.826$ being the degree of polarization of the incident neutrons and $\epsilon = 0.99$ the flipper efficiency. The polarized SANS cross sections $I(+)$ and $I(-)$ for the nanospheres and nanocubes are presented in figure 2 along with refinements according to equation (11). Refinements of the nuclear-magnetic cross term according to equation (12) converged to the same results, confirming self-consistency of the model.

Whereas previous studies on very similar nanoparticles revealed a small degree of surface spin canting along with a significantly reduced magnetization in the entire nanoparticles [14], refinements of the magnetic form factor of the ^{57}Fe-enriched nanoparticles consistently converge to a magnetic particle radius equal to the nuclear particle radius, irrespective of the starting model of either a core-shell magnetic particle or a particle with smoothly decreasing magnetization towards the surface. Hence, our analysis suggests a homogeneous magnetization distribution with spin canting throughout the entire particle without any significant degree of enhanced spin disorder close to the particle surface in either sample of the ^{57}Fe-enriched nanoparticles. Quantitative analysis of the magnetic scattering length density yields an average magnetization of 0.55(2) and 0.72(2) μ_B per iron atom for the spheres and cubes, respectively, *i.e.* significantly lower than the theoretical spin-only bulk maghemite moment of 1.25 μ_B (0 K). The room temperature maghemite moment can only be estimated, as the pure bulk material is not stable. However, the value should be close to the 0 K value, as, with $T/T_C \sim 0.3$ ($T_C \sim 900$ K[22]), the room temperature measurement is far away from the magnetic transition. Further, most theoretical models lead to values close to 1.25 μ_B, *e.g.* mean-field theory gives 1.2 μ_B and Bloch's law value is between 1.0 μ_B ($\sim T^{3/2}$ for ferromagnet) and 1.2μ_B ($\sim T^3$ for antiferromagnet)[23].

For the nanocubes, the determined magnetization is in good agreement with our earlier study on similar, but non-enriched, nanocubes of comparable particle size [14]. The smaller magnetization determined for the nanospheres is in excellent agreement with the magnetization of 0.50(5) μ_B per iron atom determined by macroscopic magnetization measurements (not shown). Our results are further in general agreement with reports on the particle-size-dependent magnetization [10, 24].

Nuclear Forward Scattering

The nuclear forward scattering data with different directions of magnetic fields are depicted in figure 3. In the data analysis, each of both crystallographic iron sites is associated with two

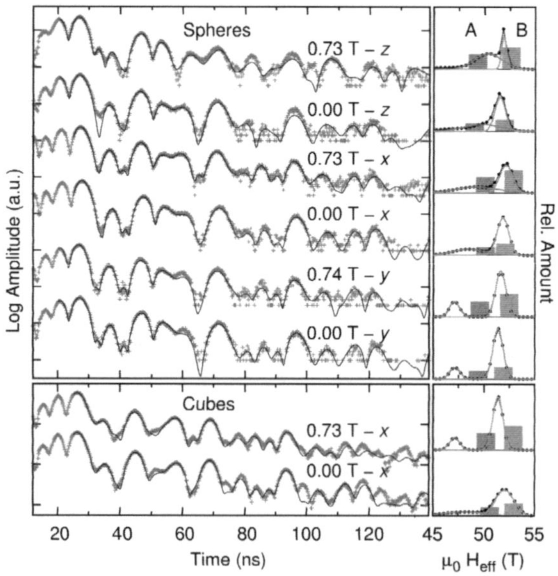

Figure 3. Adjusted NFS data on nanospheres with the indicated applied magnetic fields and a temperature of 15-30 K, depicted in logarithmic scale (left). The data were recorded at P01 with a magnetic field applied in the vertical (z) direction, horizontal direction (y), and along the beam (x). The parallel effective fields at the nuclei of A and B site (boxes) and the corresponding field distributions (Gaussian functions) are depicted for the A site in red and the B site in orange (right).

components (magnetic field along z and y) or three components (magnetic field along x). These components correspond to the relative spin orientation parallel or orthogonal to the external magnetic field. Thus, the orientation of arbitrary spins can be described with the ratio between these components, as outlined in section 3. In contrast to the introduced projection model, the hyperfine fields of the parallel and orthogonal components are not restricted to the same value. The data fitting was performed utilizing the MOTIF code[25] with fixed values of isomer shift and quadrupole interaction, taken from Mössbauer spectral results (see the supplementary data) and in agreement with literature[22, 26].

The analysis revealed that the perpendicular components of both sites are not well described with only one dedicated hyperfine field per site. Instead a Gaussian distribution of hyperfine fields is needed. One Gaussian distribution per perpendicular component was introduced, whereas the position and the width of the Gaussian are freely adjustable. The discrepancy of the fit at latter times in all spectra is probably attributed to small contributions of superparamagnetic relaxation, neglected in the analysis. Other free parameters in the adjustment are the sample thickness, an overall scaling factor, a zero shift of the measured signal, and an additional line broadening, assumed to be equal for all components.

The NFS intensities differ somewhat for different directions of the magnetic field, even for 0 T measurements. This observation is explained by different effects: first, the sample was frozen in the FC procedure leading to an alignment of the easy-axis of the nanoparticles along the field direction and thus to a preferred orientation in different directions for the individual measurements. Second, the temperatures are not exactly the same leading to slightly different hyperfine field values and also to slightly different hyperfine field distributions. For example, the temperature change between 15 and 30 K changes the hyperfine field by up to 1 T. Third, the sample thickness can slightly change between the freezing steps due to the remounting of the liquid sample, which strongly influences the time pattern.

Table 2. Adjusted parameters of the A and B site obtained from analysis of the NFS data leading to the constants $c_{\sigma\sigma}$ and $c_{\sigma\pi}$ for different magnetic fields H in direction x, y or z. The relative amount of the parallel components $\xi_{\|} = (A_{\|} + B_{\|})$ is related to these constants, as described in the theory part.

	$\mu_0 H$ (T)	$c_{\sigma\sigma}$	$c_{\sigma\pi}$	$\frac{\xi_{\|}}{\xi_{\|}+\xi_{\perp}}$
Spheres	0.73 - z	1.16(1)	-	0.46(1)
	0.00 - z	2.24(3)	-	0.31(2)
	0.73 - x	0.93(8)	0.20(4)	0.52(5)
	0.00 - x	1.61(24)	0.11(3)	0.38(6)
	0.73 - y	0.96(2)	-	0.51(1)
	0.00 - y	1.87(4)	-	0.35(1)
Cubes	0.73 - x	0.80(6)	0.28(4)	0.55(4)
	0.00 - x	1.53(16)	0.21(4)	0.39(4)

The observed differences in the spectra with different field directions seem more distinct than those in the same directions with and without applied field. However, the analysis reveals the opposite. The 0 T measurements correspond to very similar physical parameters and the application of a magnetic field changes the hyperfine field orientation and thus the ratio of aligned spins, depicted in table 2. The alignment of the spins leads to a decrease of the $c_{\sigma\sigma}$ parameter and an increase of the relative fraction of parallel spins. The value $\xi_{\|}/(\xi_{\|}+\xi_{\perp}) = 1/3$, with $\xi_{\|}$ the amount of the components with parallel hyperfine field and $\xi_{\perp}$ the amount of the components with perpendicular hyperfine field, is expected for zero field measurements, because it describes the complete 3D random case. The observed discrepancies between the obtained values at 0 T and the expected ratio of 1/3 is probably reasoned by preferred orientation of the particles due to the freezing process, remanent magnetization and time dependent macro moment reorientation, as described by Sharrock's formula[27].

In an applied field of 0.73 T, the atomic moments of the nanospheres align up to 50 %, whereas the variance is quite large with 46 − 52 %. This result is similar to magnetometry data exhibiting only a magnetization of $\sim 0.5\,\mu_B$ in contrast to the theoretically expected $1.25\,\mu_B$ for bulk maghemite (0 K). But, $\xi_{\|}/(\xi_{\|}+\xi_{\perp})$ gives only the fraction of parallel *plus* anti-parallel spins and does not indicate the relative ratio between them. This ratio between parallel and anti-parallel spins will influence the total magnetic moment significantly, as magnetometry measures the macroscopic sum of all atomic moments. The hyperfine fields and thus the spins within the larger nanocubes exhibit a stronger orientation with applied magnetic field of about 55 % as compared to the nanospheres.

The $c_{\sigma\pi}$ constant is the most difficult and the most unstable parameter to adjust. Furthermore, this component depends strongly on the applied model and can change drastically by changing the degrees of freedom in the adjustment. The strong change is probably reasoned by the frequency of the G_m term being equal to the frequency of the G_s term and thus allowing the G_m term to compensate discrepancies in the adjustment. However, the main trend of an increasing $c_{\sigma\pi}$ with application of a magnetic field and thus a non-equal orientation of spins parallel and anti-parallel along the beam is obvious. Again, comparing the $c_{\sigma\pi}$ parameter of nanospheres and nanocubes, the cubes seem to be more aligned.

In a first, rough approximation, the canting of the spins observed with NFS can be compared to the reduced magnetization obtained by SANSPOL. The canting determined by NFS relates to $0.62(6)\,\mu_B$ and $0.69(5)\,\mu_B$ for the nanospheres and nanocubes, respectively, as the magnetization is only the parallel fraction of the magnetic moments in the sample. This result is similar to the moments observed with SANSPOL, however, the nanospheres magnetic moment determined with NFS seems to be higher than observed with SANSPOL. Such difference is reasonable due to the temperature dependence of the canting[28, 29], expected to decrease the room temperature magnetization by a few percent. But this effect was not observed for the cubes, whereas the large error bars of NFS results may prohibit this observation. Another explanation for the larger moment of the nanospheres is found in $c_{\sigma\pi}$. The smaller $c_{\sigma\pi}$ of the nanospheres indicates that the relative ratio between parallel and anti-parallel spins is less pronounced, as compared to the nanocubes. Thus, a further reduction of the nanosphere moment due to compensating parallel and anti-parallel spins is possible, not considered in the rough approximation of the magnetic moments.

Summarizing, both methods, namely SANSPOL and NFS, confirm the reduced magnetization commonly observed in maghemite nanoparticles in agreement with literature and macroscopic measurements. The spatial sensitivity of the SANSPOL technique reveals a homogeneous distribution of the reduced particle magnetization, corresponding to homogeneously distributed spin disorder, with negligible effects of the particle surface. The SANSPOL results are similar to the NFS analysis, indicating a random orientation of about half of the atomic moments. Thus, roughly the half of the iron atoms are not in the perfect ferrimagnetic order.

5. Acknowledgements

SD acknowledges funding by the 7th European Community Framework Programme (PIEF-GA-2011-298918) and financial support from the Fonds der Chemischen Industrie. RPH acknowledges the Helmholtz Association of German Research Centers for funding VH NG-407 "Lattice dynamics in emerging functional materials" and support from the Materials Sciences and Engineering Division, Office of Basic Energy Sciences, U.S. Department of Energy. LB and EW acknowledge the Swedish Research Council (VR) for funding. Portions of this research were carried out at the light source PETRA III at DESY, a member of the Helmholtz Association (HGF). We would like to thank H.-C. Wille for assistance in using beamline P01.

References

[1] Kozissnik B and Dobson J 2013 *MRS Bulletin* **38** 927–932 URL http://dx.doi.org/10.1557/mrs.2013.257
[2] Pankhurst Q A, Connolly J, Jones S K and Dobson J 2003 *Journal of Physics D: Applied Physics* **36** R167 URL http://stacks.iop.org/0022-3727/36/i=13/a=201
[3] Bader S D 2006 *Rev. Mod. Phys.* **78** 1–15 URL http://link.aps.org/doi/10.1103/RevModPhys.78.1
[4] Morrish A and Sawatzky G 1971 *Proc. Int. Conf. Ferrites, University of Tokyo Press* 144
[5] Batis-Landoulsi H and Vergnon P 1983 *J Mater Sci* **18** 3399 ISSN 0022-2461 URL http://dx.doi.org/10.1007/BF00544166
[6] Coey J M D 1971 *Phys. Rev. Lett.* **27** 1140–1142 URL http://link.aps.org/doi/10.1103/PhysRevLett.27.1140
[7] Tronc E, Prené P, Jolivet J, Dormann J and Grenèche J 1998 *Hyperfine Interact* **112** 97–100 ISSN 0304-3843 URL http://dx.doi.org/10.1023/A%3A1011092712136
[8] Mørup S 2003 *Journal of Magnetism and Magnetic Materials* **266** 110–118 ISSN 0304-8853
[9] Kodama R 1999 *Journal of Magnetism and Magnetic Materials* **200** 359–372 ISSN 0304-8853 URL http://www.sciencedirect.com/science/article/pii/S0304885399003479
[10] Morales M P, Veintemillas-Verdaguer S, Montero M I, Serna C J, Roig A, Casas L, Martinez B and Sandiumenge F 1999 *Chem. Mater.* **11** 3058–3064 ISSN 0897-4756 URL http://dx.doi.org/10.1021/cm991018f
[11] Restrepo J, Labaye Y and Greneche J 2006 *Physica B: Condensed Matter* **384** 221–223 ISSN 0921-4526 URL http://www.sciencedirect.com/science/article/pii/S0921452606011434
[12] Krycka K L, Booth R A, Hogg C R, Ijiri Y, Borchers J A, Chen W C, Watson S M, Laver M, Gentile T R, Dedon L R, Harris S, Rhyne J J and Majetich S A 2010 *Phys. Rev. Lett.* **104** 207203 URL http://link.aps.org/doi/10.1103/PhysRevLett.104.207203
[13] Tronc E, Ezzir A and Cherkaoui R, Chanéac C, Noguès M, Kachkachi H, Fiorani D, Testa A M, Grenèche J M and Jolivet J P 2000 *Journal of Magnetism and Magnetic Materials* **221** 63–79 ISSN 0304-8853 URL http://www.sciencedirect.com/science/article/pii/S0304885300003693
[14] Disch S, Wetterskog E, Hermann R P, Wiedenmann A, Vainio U, Salazar-Alvarez G, Bergström L and Bückel T 2012 *New Journal of Physics* **14** 013025 URL http://stacks.iop.org/1367-2630/14/i=1/a=013025
[15] Park J, An K, Hwang Y, Park J G, Noh H J, Kim J Y, Park J H, Hwang N M and Hyeon T 2004 *Nat Mater* **3** 891 ISSN 1476-1122 URL http://dx.doi.org/10.1038/nmat1251
[16] Ahniyaz A, Sakamoto Y and Bergström L 2007 *Proceedings of the National Academy of Sciences* **104** 17570–17574 URL http://www.pnas.org/content/104/45/17570.abstract
[17] Wetterskog E, Agthe M, Mayence A, Grins J, Wang D, Rana S, Ahniyaz A, Salazar-Alvarez G and Bergström L 2014 *Science and Technology of Advanced Materials* **15** 055010– ISSN 1468-6996 URL http://stacks.iop.org/1468-6996/15/i=5/a=055010
[18] Tarr B R, Booth H S and Dolance A 1950 *Inorganic Syntheses* (John Wiley and Sons, Inc.) pp 191–194 URL http://dx.doi.org/10.1002/9780470132340.ch51
[19] Dewhurst C 2008 Graphical reduction and analysis sans program for matlab URL https://www.ill.eu/instruments-support/instruments-groups/groups/lss/grasp
[20] Dattagupta S 1981 *Hyperfine Interact* **11** 77–126 ISSN 0304-3843 URL http://dx.doi.org/10.1007/BF01026470
[21] Shvyd'ko Y V 1999 *Hyperfine Interact* **123-124** 275–299 ISSN 0304-3843 URL http://dx.doi.org/10.1023/A%3A1017015721916
[22] Long G J 1987 *Mössbauer Spectroscopy Applied to Inorganic Chemistry Volume 2* (Plenum Press)
[23] Blundell S 2001 *Magnetism in Condensed Matter* (Oxford)
[24] Millan A, Urtizberea A, Silva N, Palacio F, Amaral V, Snoeck E and Serin V 2007 *Journal of Magnetism and Magnetic Materials* **312** L5–L9 ISSN 0304-8853 URL http://www.sciencedirect.com/science/article/pii/S0304885306010717
[25] Shvyd'ko Y V 2000 *Hyperfine Interact* **125** 173–188 ISSN 0304-3843 URL http://dx.doi.org/10.1023/A%3A1012633620524
[26] Helgason O, Rasmussen H K and Morup S 2006 *Journal of Magnetism and Magnetic Materials* **302** 413–420 ISSN 0304-8853 URL http://www.sciencedirect.com/science/article/pii/S0304885305007973
[27] Bedanta S, Petracic O and Kleemann W 2014 *Handbook Of Magnetism Vol. 22* (in print)
[28] Zulfiqar, Rahman M, Usman M, Hasanain S, ur Rahman Z, Ullah A and Kim I 2014 **65** 1925–1929– ISSN 0374-4884 URL http://dx.doi.org/10.3938/jkps.65.1925
[29] Guivar J, Martinez A, Anaya A, Valladares L, Felix L and Dominguez A 2014 *Advances in Nanoparticles* **3** 8 URL http://www.scirp.org/journal/PaperInformation.aspx?PaperID=49326

6. Supplementary data

Mössbauer spectroscopy was measured on the enriched spheres and cubes at 10 K without applied magnetic field, see figure 4. The spectra are calibrated with respect to α-Fe. The fit model assumes two crystallographic sites and a Gauss-distributed hyperfine field. The ratio of the A and B sites are fixed to 1:1.75, in agreement with the NFS data. The obtained hyperfine parameters are given in table 3. The maghemite spectra can contain different iron-oxide impurities, likely wuestite. These impurities usually form additional peaks at the indicated energies (arrows). For the samples used in this manuscript, such impurities are not adjustable in the Mössbauer spectra.

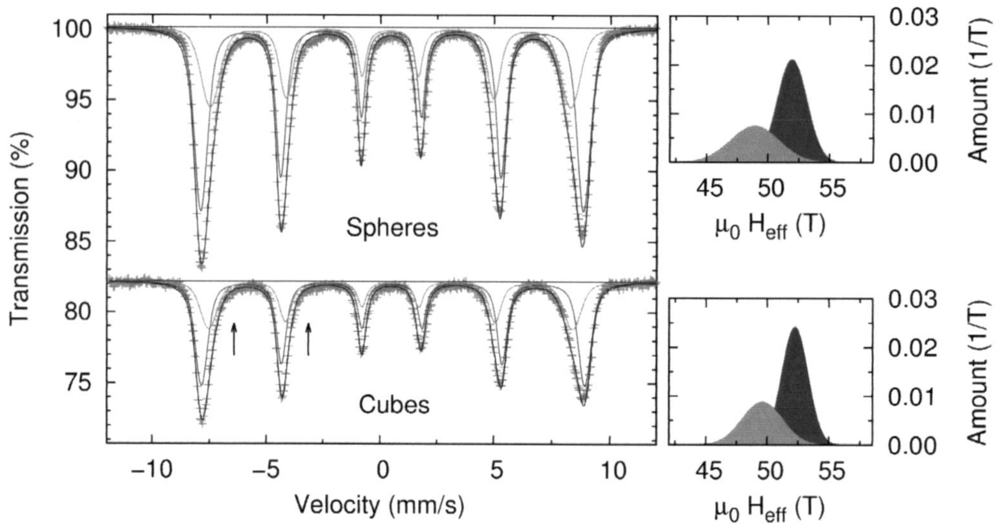

Figure 4. Adjusted Mössbauer data on nanospheres and nanocubes (left) with the corresponding hyperfine field distribution (right). The ratio of the maghemite A site (pink) and B site (blue) are fixed to 1:1.75, in agreement with the NFS data. Possible impurities, likely other oxides like wuestite, commonly lead to peaks at the indicated energies (arrows) and their presence can be excluded at the $\sim 1\,\%$ level.

Table 3. Obtained parameters from Mössbauer spectroscopy. The isomer shift δ, the quadrupole interaction ΔE_Q, the center value of the Gaussian field distribution H^{center}, and the standard deviation of the Gaussian distribution σ are given for the A and B site of both samples.

	site	δ (mm/s)	ΔE_Q (mm/s)	$\mu_0 H^{center}$ (T)	σ (T)
Spheres	A	0.40(1)	-0.03(1)	49.0(1)	2.0(3)
	B	0.48(1)	0.00(1)	52.0(1)	1.2(1)
Cubes	A	0.39(1)	-0.04(2)	49.6(1)	1.7(3)
	B	0.48(1)	0.00(1)	52.3(1)	1.1(1)

Structure of hydrogenous liquids: separation of coherent and incoherent cross sections using polarised neutrons

A Stunault[1], S Vial[1], L Pusztai[2], G J Cuello[1] and L Temleitner[2]

[1] Institut Laue-Langevin, 71 Av. des Martyrs, CS 20156, F-38042, Grenoble, France
[2] Institute for Solid State Physics and Optics, Wigner Centre for Physics, Hungarian Academy of Science, Budapest, Hungary

E-mail: stunault@ill.fr

Abstract. The determination of the coherent structure factor of hydrogenous liquids is very difficult: while X-rays are barely sensitive to hydrogen, neutrons results still lack accuracy due to the contamination of the scattering intensities by a huge spin-incoherent signal from the ^{1}H atoms. Using polarised neutrons with polarisation analysis, one can experimentally separate the coherent and incoherent contributions to the scattered intensity. We present the upgrade of the D3 polarised hot neutron diffractometer at ILL to study hydrogenated liquids. We show first data obtained from a test sample of water and detail the data reduction leading to an unprecedented accuracy in the extraction of the coherent signal, representative of the structure.

1. Introduction

The determination of the coherent structure factor of hydrogenous liquids by the usual scattering methods is very difficult: the scattering of X-rays by 1-electron hydrogen is usually too weak to be detected, while neutron scattering is largely dominated by incoherent scattering (incoherent cross-section $\sigma_i = 80.27(6)$ barns, to compare to the coherent cross-section σ_c=1.7523(10) barns [1]). As a result, the parasitic incoherent background may reach over 90 % of the measured signal, as seen *e.g.* in the case of light water H_2O [2][3] . Theoretical efforts to estimate the incoherent neutron background have not yet led to a unique solution and, as a result, the precise nature of the hydrogen bond is still highly debated.

In the case of isotopic ^{1}H hydrogen, the incoherent scattering is solely due to the disorder of the nuclear spins ("spin-incoherent"). It can be shown that in that case, the scattering of the neutrons partially (2/3) reverses their polarisation ("spin-flip", SF) , while coherent and isotope incoherent scattering do not ("non-spin-flip", NSF) [4]. As a result, the coherent and incoherent intensities are linear combinations of the spin-flip and non-spin-flip measured intensities, Eqs. (1) and (2).

$$I_{coh}(Q) + I_{isotope\_inc}(Q) = I_{NSF}(Q) - \frac{1}{2} I_{SF}(Q) \quad (1)$$

$$I_{spin\_inc}(Q) = \frac{3}{2} I_{SF}(Q) \quad (2)$$

Using polarised neutrons with polarisation analysis, one can then work around the high-background problem and experimentally separate the coherent and incoherent contributions to the scattered intensity. Note that Eqs. (1) and (2) only take the elastic scattering (coherent and incoherent) into account. Corrections for inelasticity, multiple scattering... are beyond the scope of this paper.

2. Instrument

Although the above described principle is quite simple, very few structural studies of hydrogenated liquids using this technique have as yet been reported [5][6]. The main reason lies in a former shortcomings of the experimental technique: studies of liquids need measurements up to high momentum transfers Q (typically 20-25 Å) for the purpose of Fourier transforming [7], hence need the use of neutrons with a relatively short wavelength, typically 0.4-0.5 Å. At these wavelengths, the use of supermirrors for polarisation analysis is excluded (very small critical angle), and the use of polarising crystal analysers is limited (small take off angles, implying the need of either forbiddingly large crystals, or very long distances. The situation has changed with the advent of spin filters based on polarised ^{3}He [8] and the ensuing developments allowing their easy use even at short neutron wavelengths [9].

The D3 diffractometer of the Institut Laue-Langevin (ILL) in Grenoble, initially built for magnetic (elastic) studies on single crystals using polarised neutrons has been upgraded for the study of liquids (Fig. 1). The neutron beam on D3 comes from the hot source and the covered wavelength range is 0.4 Å to 0.84 Å. The polarising monochromator is a Heusler, Cu_2MnAl, single crystal. Guide fields along the beam path ensure the preservation of the neutron polarisation. A monitor before the sample serves for normalisation of the measured intensities. The neutron polarisation can be flipped either before or after the sample using a nutator. The ^{3}He spin filter is situated just before the single ^{3}He detector, on the detector arm. The detector is scanned over a total angular range of 4° to 120°, with typical data acquisition times of 1-2 minutes/angular step, both polarisation channels being measured at each step. Using a wavelength of 0.52 Å, the covered Q-range is 0.8 to 21 Å^{-1}. With a beam size of 12 x 22 mm^2 and an annular vanadium sample cell, with a 1 mm thick annular space (to minimise multiple scattering), the amount of sample in the beam is about 0.9 cm^3.

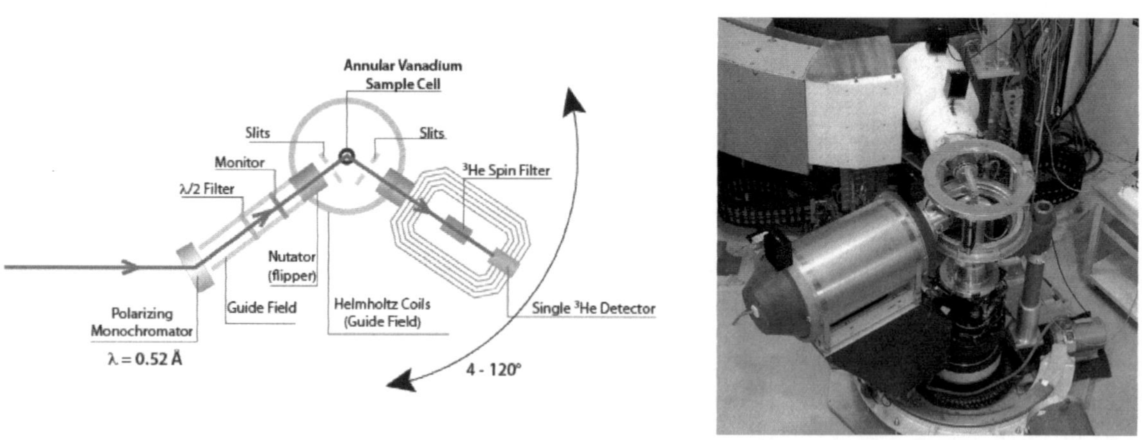

Figure 1. The D3 diffractometer at ILL, modified for the study of liquids: principle (left) and top view (right).

Since the ^{3}He polarisation in the spin filter decays over time, typically loosing about 5-10% per day, the analysing power has to be monitored. For this purpose, a silicon single crystal is glued to the top of the sample cell and regularly driven into the beam via a motorised translation. The polarisation of the beam is measured by comparing the SF and NSF intensities of a Si Bragg peak.

In the instrument upgrade, particular care has been taken in the shielding to minimise the instrument background. Typical levels away from the vicinity of the direct beam are of the order of 1 count/s.

3. Data reduction

The first step of the data reduction is the accurate determination of the spin-filter efficiency, and its relaxation rate. Following [8], the transmissions of the up T^+ and down T^- neutron polarisations are:

$$T^{\pm} = \frac{1}{2} e^{-O(\lambda)(1 \mp Pol_{He})}, \tag{3}$$

where $O(\lambda)$ is the opacity, proportional to the ^{3}He pressure P, the length of the spin filter l and the wavelength λ. At 295 K, $O(\lambda) = 7.28210^{-2} P[bar] l[cm] \lambda[\AA]$. Pol_{He} is the ^{3}He polarisation, $Pol_{He} = Pol^0_{He} e^{-t/T_1}$, where t is the time of the measurement, Pol^0_{He} the ^{3}He polarisation at t = 0, and T_1 is the relaxation time. The polarisation measurements on a Bragg peak actually give access to the product of the initial beam polarisation and the spin-filter analysing power (efficiency). Fig. 2 shows a typical result. Currently on D3, T_1 is of the order of 100 to 120 hours, and the polarisation of the small beam impinging on the single crystal used for calibration is usually of the order of 90%. We must note at this point that, since the single crystal used for calibration of the polarisation is much smaller than the sample cell, we also measure the full beam polarisation using a larger Si wafer to correct for a possible different overall polarisation of the full beam.

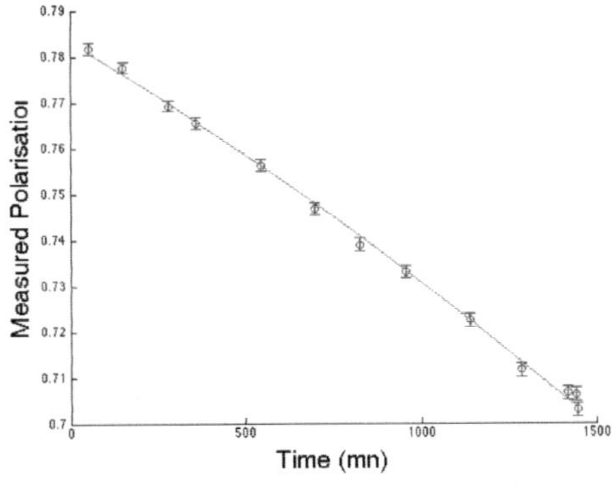

Figure 2. Calibration of the spin filter analysing power: fit of the relaxation time T_1 and the incoming beam polarisation

Figure 3 shows the different steps of the data reduction for a light water H_2O sample, from the top left panel (measurement) to the bottom right panel (coherent and incoherent cross sections). The first step (top right) is the correction for the decay of the spin filter analysing power. Once this critical step has been performed, one can average the corrected scans, hence improving the statistics (bottom left), and extract the coherent and incoherent cross sections (bottom right). In the figure, the coherent part, representative of the correlations, had to be blown up by one order of magnitude for visibility (see the different scales). A measurement with unpolarised beam, as in [2][3] is actually the sum of both spectra and, as pointed out in the introduction, widely dominated by the incoherent signal.

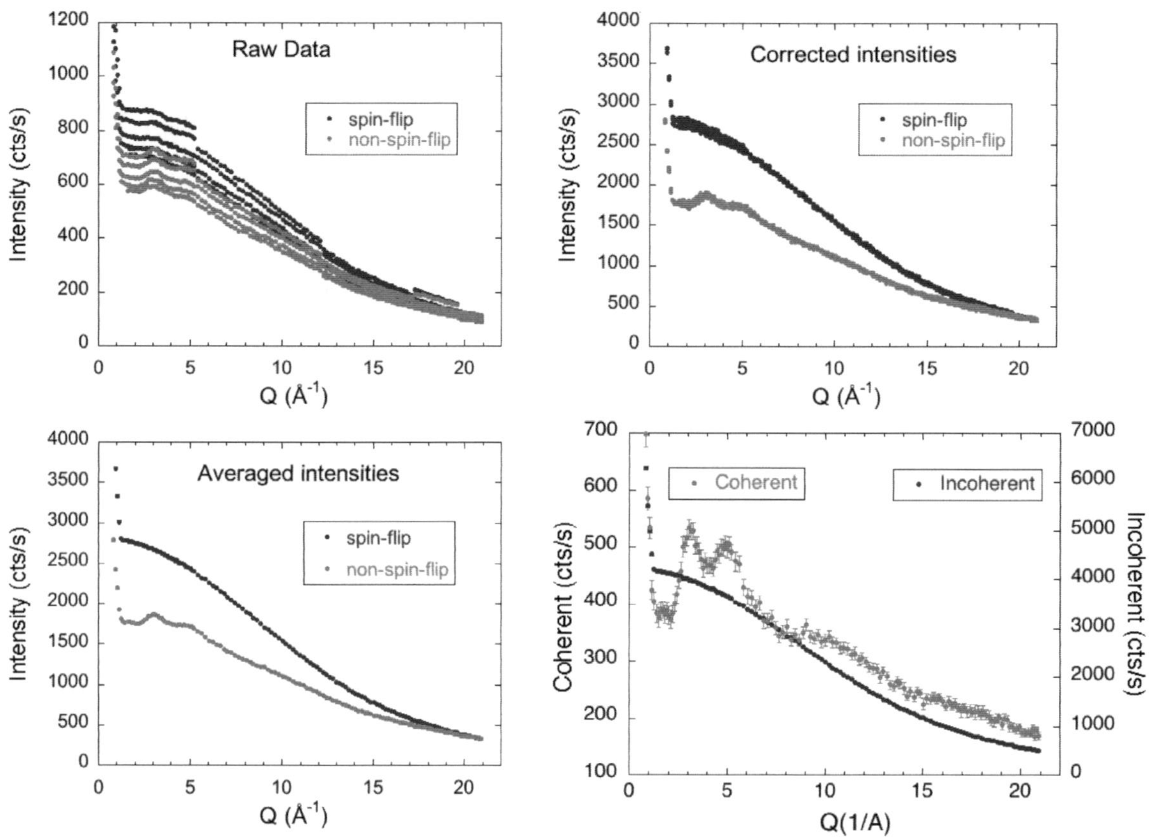

Figure 3. Data reduction in the case of H_2O at room temperature. The spread of the raw data (top left panel) and the observed discontinuities reflect the time dependence of the spin filter efficiency, since, for practical reasons, the scans were not run in a continuous way.

4. Discussion

Obviously, the uncorrected incoherent and coherent data in Fig. 3, while showing the power of the method, should not be used to extract correlation functions without a more careful treatment of the NSF and SF corrected and averaged intensities. For example, one must take into account the fact that the thin container is made of Vanadium, which is a mostly spin-incoherent scatterer, giving signal in both polarisation channels following the "1/3 - 2/3" rule from Eqs. (1) and (2). As usual in that type of measurement, a more tricky correction comes from multiple scattering, for which the usual treatments have to be adapted to take the neutron spin state into account (scattering twice in the SF channel will actually give intensity in the NSF channel...). Details on the current state of the analysis can be found in [10]. Another approach is to use the polarised neutron results to model the incoherent contribution to the scattering [11], and use this knowledge in the analysis of unpolarised data.

5. Conclusion

The D3 diffractometer of the ILL is now fully operational for the study of liquids using polarised neutrons, and scripts have been developed for on-line data reduction. The data acquisition is currently slower than on fully dedicated instruments like D4 at ILL [12] due to the absence of a wide-angle spin filter and a multi-detector. Typical measuring times are 1-2 days per sample, which is fortunately not forbiddingly long. Using D3 for accuracy in the case of the non-deuterated samples, and complementing with unpolarised results for more deuterated samples may be an efficient option. A future upgrade will be the installation of a vacuum chamber to further minimise the instrument background, especially in the vicinity of the direct beam and allow measurements down to lower Q values. As an ultimate upgrade, a wide-angle spin-filter for short wavelengths should be developed, in association with the use of a multidetector.

Acknowledgments

This research was performed with the approval and support of the Institut Laue-Langevin, DOI:10.5291/ILL-DATA.6-02-536

References

[1] V F Sears 2007 *Neutron News* **3** 26.
[2] W E Thiessen and A H Narten 1982 *J. Chem. Phys.* **77** 2656.
[3] A K Soper 2013 *ISRN Physical Chemistry* 2013 (Hindawi Publishing Corporation), Article ID 279463, [http://dx.doi.org/10.1155/2013/279463], and included references.
[4] R M Moon, T Riste and W C Koehler 1969 *Phys. Rev.* **181** 920.
[5] J C Dore, J H Clarke and J T Wenzel 1976 *Nucl. Instr. Meth.* **138** 317.
[6] L Temleitner, L Pusztai and W Schweika, *J. Phys.: Cond. Matter* **19**, 335207.
[7] G J Cuello 2008 *J. Phys.: Cond. Matter* **20** 244109.
[8] K P Coulter, T E Chupp, A B McDonald, C D Bowman, J D Bowman, J J Szymanski, V Yuan, G D Cates, D R Benton and E D Earle 1990 *Nucl. Instr. Meth. A* **288** 463.
[9] K H Andersen, D Jullien, A K Petoukhov, P Mouveau, F Bordenave, F Thomas and E Babcock 2009 *Physica B* **404** 2652.
[10] L A Rodriguez Palomino, G J Cuello, J Dawidowski, L Temleitner, L Pusztai and A Stunault 2015, *J. Phys.: Conf. Series* **663** 012002.
[11] L Temleitner, A Stunault, G J Cuello and L Pusztai 2015, *Phys. Rev, B* **92** 014201.
[12] H E Fischer, G J Cuello, P Palleau, D Feltin, A C Barnes, Y S Badyal and J M Simonson 2002 *Applied Physics A* **74** S160

The study of perpendicular magnetic anisotropy in the magnetic sensors with linear sensitivity using polarized neutron reflectometry

T. Zhu

China Spallation Neutron Source, Institute of High Energy Physics, Chinese Academy of Sciences, Dongguan 523803, P.R. China
Institute of Physics, Chinese Academy of Sciences, Beijing 100190, P. R. China

E-mail: tzhu@aphy.iphy.ac.cn

Abstract. The CoFeB sandwiched by Ta and MgO layers enables a perpendicular magnetic anisotropy (PMA) and provides a pathway for such application. In this paper, we reported the origin of PMA in CoFeB using the anomalous Hall effect (AHE) and polarized neutron reflectometry (PNR). From PNR experiments, we obtained the details of the magnetic and structural depth profiles inside the film. It is found that the PMA properties of CoFeB layers deposited above and under MgO layer are different and PNR measurements confirmed that a large PMA in the CoFeB above MgO layer is related to its low magnetization. Based on this PMA mechanism, we obtain a high sensitivity of AHE in the perpendicular CoFeB, which opens a new avenue to detect ultralow magnetic field.

1. Introduction

The magnetic sensors for high sensitive magnetic detecting have been reported using magnetoresistive effect or anomalous Hall effect, where a linear and reversible response is produced [1-5]. For example, detection of specific biomolecular interactions through sensing of nanoscopic magnetic labels provides one of the most promising routes towards biosensing with solid-state devices [3-5]. Typically, these magnetic nanoparticles are superparamagnetic or non-remanent ferromagnetic particles and can be attached to target biomolecules. Under a small magnetic field, these particles acquire a magnetic moment which produces a fringe field over the sensor [3]. In each probe, a magnetic sensor or Hall sensor detects the stray field produced by the label giving an electrical signal [3,4]. There are several methods to detect such weak remanent magnetic field. The sensors for biosensing have been reported using magnetoresistive effect [3]. By using a complicated design with an orthogonal alignment between the sensing and reference magnetization, a linear and reversible giant magnetoresistance or tunneling magnetoresistance response is produced [6]. On the other hand, micro-Hall magnetic sensors based on semiconductor heterostructures have been attracting wide interests [4,5]. As a simple solution, the linear anomalous Hall effect (AHE) exhibits fast response, high sensitivity, and ease of integration due to the cheap and simple fabrication process [1,2]. The Hall resistance in a magnetic thin film consists of two terms: one presents the ordinary Hall effect and another presents the AHE which is much larger than the former one [7]. The sensitivity of linear anomalous Hall effect (S_{AHE}) is defined as the ratio of anomalous Hall resistance (R_{AH}) to the out-of-plane saturation field (H_S),

R_{AH}/H_S. So far, large SAHE has been observed in Co(Fe)/Pt multilayers or granular films [8,9], due to the induced perpendicular magnetic anisotropy in these thin films.

Recently, the magnetization of CoFeB has been reported to stay out of plane when it is sandwiched by MgO and Ta layers [10-12]. In this paper, we reported that the effective magnetic anisotropy of CoFeB layer in the MgO/CoFeB/Ta thin film depended on the layer deposition sequences using the polarized neutron reflectometry (PNR). And then, we demonstrate a giant linear anomalous Hall effect in the MgO/CoFeB/Ta thin film, which detects the weak magnetic field from a superparamagnetic label.

2. Method

2.1. Sample preparation

MgO (1-5)/CoFeB (0.8-3)/Ta (0.55-5) samples were deposited on a thermally oxidized silicon wafer by magnetron sputtering with a base pressure of 5×10^{-6} Pa. The numbers in brackets are nominal thicknesses in nanometers. Ar was used as the sputtering gas and the sputtering pressure for metals was 0.4 Pa. MgO layers was deposited by radio-frequency (RF) sputtering and the sputtering pressure was 0.2 Pa. High purity of Ta (99.95%), MgO (99.99%), and $Co_{40}Fe_{40}B_{20}$ (99.9%, Functional Materials International, Japan) were used as the target materials. All the samples were capped with a 3-nm-thick MgO to prevent oxidation. The layer thickness was identified by an x-ray reflectometer (XRR, Brucker D8).

2.2. Anomalous Hall effect measurements

The patterned Hall bar samples were annealed from 90 to 450°C for 1 hour. Room temperature anomalous Hall resistivity, ρ_{AHE}, and longitudinal resistivity, ρ_{xx}, were measured in a commercial Hall probing system by using an electromagnet (up to 1.5 T) and high resolution multimeters (Keithley@ 220 and 2182). To thoroughly eliminate the ordinary Hall effect, R_{AH} is defined as the zero field extrapolation of the high field data. The low temperature AHE measurements were carried out by using standard cross-structure four-point setting in a physical property measurement system (PPMS, Quantum Design@ PPMS-14T).

2.3. Polarized neutron reflectivity measurements

The PNR technique was used to directly measure the depth-dependent magnetization profile of the samples to investigate the contribution of the demagnetization term to PMA. Room-temperature PNR measurements were carried out on the Magnetism Reflectometer (BL-4A) at the Spallation Neutron Source at Oak Ridge National Laboratory [13].

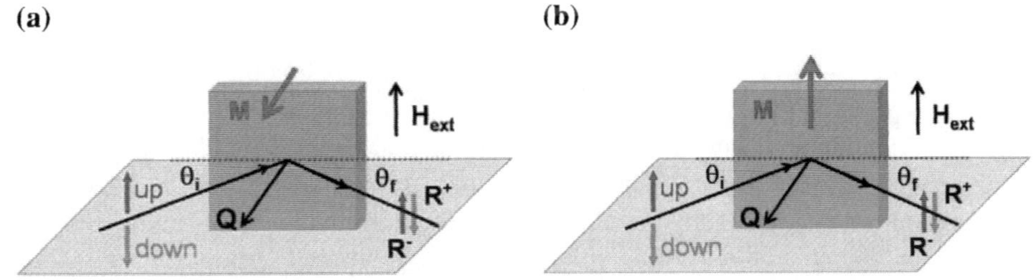

Figure 1. The schematics of the PNR measurement for a perpendicular thin film under (a) a low and (b) a high external fields.

In general, PNR is a technique sensitive to the compositional and in-plane magnetic depth profiles of thin films but not sensitive to the layer magnetization perpendicular to the sample surface, because the magnetic neutron scattering cross section is sensitive only to components of $\mathbf{M} \perp \mathbf{Q}$ [14]. Q =

$4\pi\sin\theta/\lambda$, where λ is the neutron wavelength and θ the incident angle of the neutron beam, respectively. Hence, it needs a high external field to saturate the film perpendicular to **Q**. Figure 1 shows the schematics of the PNR measurement for a perpendicular thin film. There is no in-plane component of **M** under the low external field (figure 1(a)). The reflectivity curves look like unpolarized neutron reflectivity. On the other hand, the magnetization will align with the external field when it is large enough (figure 1(b)), and then the magnetic depth profiles of thin films have been obtained.

3. Results and Discussions

3.1. Perpendicular magnetic anisotropy

Figure 2 shows the CoFeB thickness dependence of ρ_{AHE} and ρ_{xx} for annealed MgO (1.1)/CoFeB (t)/Ta (2.2) thin films (annealing temperature T_a = 300°C). The insets show the representative Hall resistance hysteresis loops (R_{xy}-H) with t_{CoFeB}. Similar with previous results [10,11], obvious perpendicular magnetic anisotropy has been obtained when CoFeB thickness ranging from 1.06 to 1.5 nm in the Ta capping films.

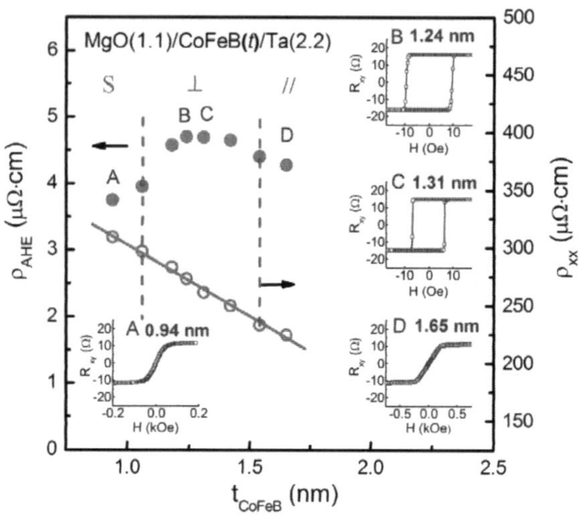

Figure 2. CoFeB thickness dependence of ρ_{AHE} (the solid circles) and ρ_{xx} (the open circles) for annealed MgO (1.1)/ CoFeB (t)/Ta (2.2) thin films (T_a = 300°C) with S defined as super-paramagentic, ⊥ as magnetic easy axis perpendicular to film, and // as magnetic easy axis in the plane. The insets show the representative Hall resistance hysteresis loops.

Furthermore, ρ_{AHE} of the samples are larger than those of ordinary transition metals (~1 μΩ·cm [7]). As for longitudinal resistivity, ρ_{xx} of all the annealed samples is several hundred μΩ·cm and increases with the decrease of CoFeB thickness. The resistivity of tantalum is about 200 μΩ·cm due to its extremely short mean free path (1 nm) [15]. Meanwhile, the resistivity of CoFeB is also much higher (about 160 μΩ·cm) [16]. Due to these layers with high resistivity, MgO/CoFeB/Ta thin film has large ρ_{xx}, which is much larger than those for a traditional Pt based multilayer or alloy. The large ρ_{xx} itself is very important, because low ρ_{xx} results in a blockade to reduce the device size [17].

The sample for PNR measurement was Ta (2.2)/CoFeB (1.04)/MgO (1.1)/ CoFeB (1.22)/Ta (2.2) (T_a = 300°C), hereafter defined as Ta/CoFeB/MgO/CoFeB/Ta, with the sample size of 10×15 mm^2. Figure 3(a) shows the R_{xy}-H loop of the sample. Apparently, the PMA properties of CoFeB layers deposited above and under MgO layer are different. For comparison, two samples Ta (2.2)/CoFeB (1.04)/MgO (1.1) and MgO (1.1)/CoFeB (1.22)/Ta (2.2), hereafter defined as MgO/CoFeB/Ta and Ta/CoFeB/MgO, respectively, were prepared on Si/SiO$_2$ (T_a=300°C). As seen from Fig. 3(b), the coercivity field (H_C) of MgO/CoFeB/Ta is nearly 100 Oe, which is about 24 Oe in Ta/CoFeB/MgO, which is consistent with the H_C of the related CoFeB layers in the sample for the PNR measurement, *i.e.* the CoFeB above or below the MgO, as shown in Fig. 3(a). The results also confirm that the PMA properties of CoFeB are dependent on the stack deposition sequence.

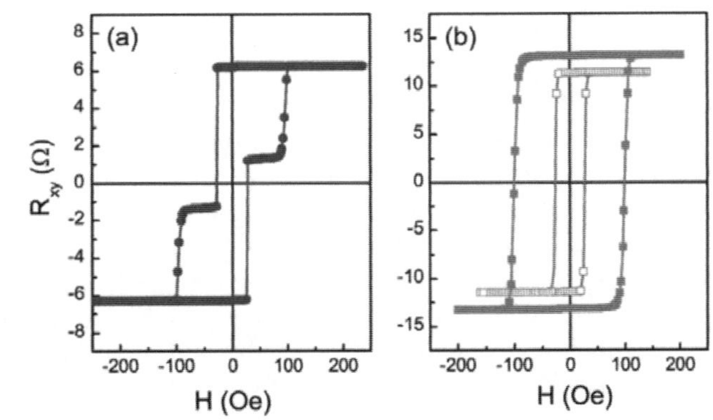

Figure 3. Hall resistance hysteresis loops of (a) Ta(2.2)/ CoFeB(1.04)/MgO(1.1)/CoFeB(1.22)/Ta(2.2) (blue solid circles) and (b) MgO(1.1)/ CoFeB(1.22)/Ta(2.2) (red solid squares) and Ta(2.2)/CoFeB(1.04)/MgO(1.1) (red open squares) films annealed at 300 °C.

3.2. Polarized neutron reflectivity

In a reflectometry experiment, the interaction of neutrons with a magnetic film is described by the Fermi pseudopotential $V_{\pm}$,

$$V_{\pm} = \frac{2\pi\hbar}{m} N(b_n \pm b_m), \tag{1}$$

where m denotes the neutron mass, N the atomic number density, b_n the nuclear scattering length, and b_m the magnetic scattering length, respectively [18]. For the two neutron spin states, the magnetic scattering length adds to or subtracts from the nuclear one [19], leading to different momentum transfer Q dependence of reflectivity curves, R^+ and R^-, and is used to determine the magnetic moment of magnetic films. Here, R^+ denotes the reflectivity for spin-up neutrons with spin parallel to H_{ext}, and R^- denotes the reflectivity for spin-down neutrons with spin antiparallel to H_{ext}. The sample's depth-dependent nuclear scattering length density Nb_n and magnetic scattering density Nb_m, hereinafter defined as $SLD_N(z)$ and $SLD_M(z)$, respectively, and then the magnetization are determined by a simultaneous fit of the PNR data.

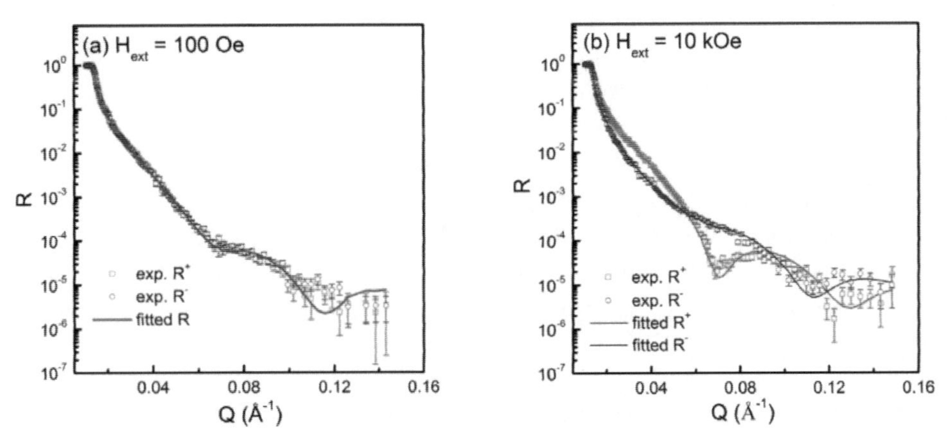

Figure 4. The experimental R^+ (open squares) and R^- (open circles) measured at room temperature and (a) H_{ext}= 100 Oe and (b) H_{ext}= 10 kOe as functions of Q, respectively. The solid lines are the fitted theoretical curves.

Figure 4(a) shows R^+ (open squares) and R^- (open circles) as a function of Q measured in an external field of 100 Oe after the saturation. This external field is much smaller than the out-of-plane anisotropy field. In such low external field, the magnetization **M** is perpendicular to the plane of the sample and parallel to **Q**. Meanwhile, the small external magnetic field $\mathbf{H}_{\mathbf{ext}}$ provides a guide field for the polarized neutron beam. This is confirmed by the PNR data, when the reflectivities for the two spin-states, R^+ and R^-, overlap. Thus R^+ and R^- have the same scattering vector Q dependence when H_{ext} is much smaller than the out-of-plane anisotropy field of a PMA film.

To determine the magnetization profile in the sample, a large in-plane field is applied to saturate the out-of-plane magnetization into the hard in-plane direction. Figure 4(b) shows the PNR data measured at room temperature with H_{ext} = 10 kOe, which is sufficient to obtain the in-plane saturation of the magnetization. This is to say, **M** is in the plane of the sample and perpendicular to **Q** when H_{ext} = 10 kOe.

Figure 5 shows the structural and magnetic profiles obtained from a fit to the data. First, we fitted 100 Oe PNR data to get the parameters of the layer structure as seen in the left and middle panels of figure 5. Second, we fixed these structural parameters to fit the 10 kOe PNR data to get the magnetization of the individual layer of the sample as seen in the right panel of figure 5.

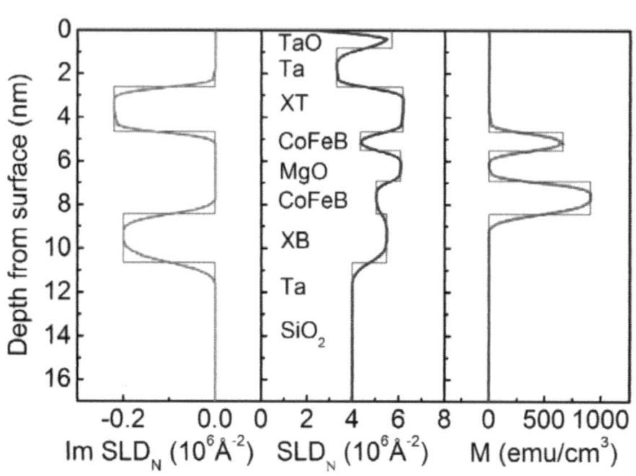

Figure 5. The depth-dependence of Im-SLD$_N(z)$ (left), SLD$_N(z)$ (middle), and magnetization of CoFeB layer from SLD$_M(z)$ (right). The thin green lines show the thicknesses of layers in the sample obtained from the fit to the data

In the CoFeB sandwiched by MgO and tantalum layers, the effective magnetic anisotropy can be generally defined as

$$K_{eff} = (K_b - 2\pi M_S^2) + 2K_S / t_{CoFeB}, \quad (2)$$

where t_{CoFeB} is the CoFeB thickness and M_S is the saturation magnetization of the CoFeB layer. K_b is the bulk crystalline anisotropy, which can be neglected [10]. Therefore, the perpendicular anisotropy in the CoFeB is the result of the competition between the demagnetization energy ($2\pi M_S^2$) and the interfacial anisotropy energy (K_S/t_{CoFeB}). One pronounced feature from the PNR results is that the magnetization of the CoFeB layer above MgO ($M_{S,above}$ = 663 emu/cm^3) is significantly smaller than that of the CoFeB layer under MgO ($M_{S,under}$ = 906 emu/cm^3). According to Eq. 2, the smaller demagnetization term ($2\pi M_{S,above}^2$) will induce a larger positive K_{eff} and a larger switching field. On the other hand, the demagnetization term of CoFeB under MgO ($2\pi M_{S,under}^2$) is much larger, leading to a dramatic decrease in K_{eff}.

The mechanism of the difference in magnetization between CoFeB layers above and below MgO layers remains unclear, but it may be related to the structural dependent formation of magnetically dead layer (MDL) at the CoFeB-Ta interface during deposition, or intermixing upon annealing, or the roughness of the underlayer on which CoFeB layer was deposited [20]. In general, a reduction of the magnetization in a thin magnetic film is normally believed to be due to the presence of a MDL at the interface. For CoFeB-Ta interface, there is an approximately 0.5-nm-thick MDL in the case of the deposition of CoFeB above MgO (*i.e.* MgO/CoFeB/Ta) but no MDL in the case of the deposition of CoFeB under MgO (*i.e.* Ta/CoFeB/MgO) [10,21,22]. Based on the PNR fitting results, the SLD_N of CoFeB above MgO is 4.35×10^{-6} Å^{-2}, whereas the SLD_N of CoFeB under MgO is 5.03×10^{-6} Å^{-2}. It suggests the quality of CoFeB layers is one of possibilities for the magnetization reduction besides of the effect of MDL. The observed low magnetization of the CoFeB layer above MgO may be due to the formation of a loose structure in the CoFeB layer during the deposition on MgO.

Another pronounced feature from the PNR results is the observation of boron diffusion after annealing. The exceptional sensitivity of neutrons to boron because of its absorption cross section is a distinct advantage and allows accurate determination of the depth profile of the boron distribution. As shown in Fig. 5, the interface layers between the CoFeB and tantalum layers are 2 nm (near top of sample) and 2.2 nm (near bottom of sample), hereinafter defined as XT and XB, respectively. The SLD_N of XT is $6.1721\times10^{-6} + 2.2\times10^{-7}i$ Å^{-2}, with the second term defined as Im-SLD_N from the absorption cross section, whereas the SLD_N of XB is $5.4878\times10^{-6} + 2\times10^{-7}i$ Å^{-2}. The nominal SLD of boron is $6.94\times10^{-6} + 2.79\times10^{-7}i$ Å^{-2} [23]. From the fit to the PNR data, we observed that after the annealing, most of the boron diffused from the CoFeB layers where it was initially found to form 2-nm-thick interface layers, which are a mixture of boron and tantalum.

3.3. Linear anomalous Hall effect

Figure 6 shows the representative R_{xy} - H loop of MgO (2.5)/CoFeB (1.2)/Ta (0.7)/MgO (3) thin film annealed at 210 °C. The magnetization turns to out-of-plane when H reaching a small field, which is the saturated field H_S. Beyond the H_S, the anomalous Hall resistance accesses its saturation soon. And then a large linear R_{xy} - H loop is observed. The S_{AHE} at room temperature is 2376 Ω/kOe when t_{Ta} = 0.7 nm, which is 21 times larger than that for the best semiconductor (110 Ω/kOe for InSb [5]).

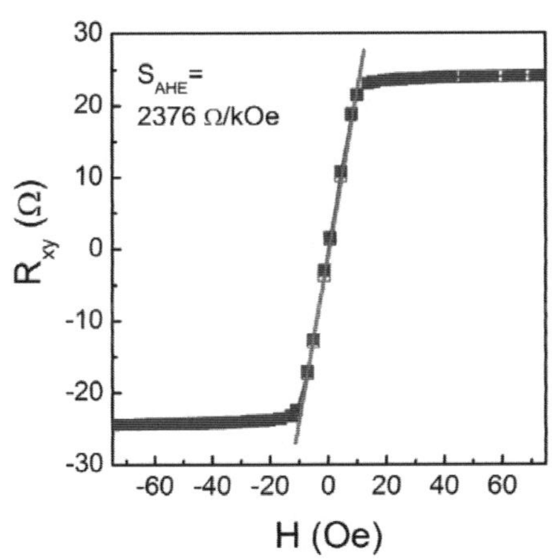

Figure 6. The representative curves of R_{xy} -H for MgO (2.5)/CoFeB (1.2)/Ta (0.7)/ MgO (3) film (T_a = 210°C). The red solid line is guide for eyes.

Besides that obvious PMA has been obtained when the CoFeB thickness ranging from 1.06 to 1.5 nm in the MgO/CoFeB/Ta thin films annealed at 300°C, linear R_{xy} - H curves have been observed in the films with thin CoFeB layers ($t_{CoFeB} \leq 1$ nm) and thick CoFeB layers ($t_{CoFeB} > 1.5$ nm) as shown in the insets of Fig. 2. It should be pointed out that those two magnetization states are different. The film with thin CoFeB layer shows out-of-plane superparamagnetic at room temperature, because out-of-plane ferromagnetic returns at low temperature. On the other hand, the film with thick CoFeB layer shows in-plane ferromagnetic state when $t_{CoFeB} > 1.5$ nm.

According to Eq. 2, K_{eff} is the result of the competition between the demagnetization energy ($2\pi M_S^2 t_{CoFeB}$) and the interfacial anisotropy energy. The magnetization of thin film shows out-of-plane when $K_{eff} > 0$. From the PNR measurement, the PMA in CoFeB above MgO layer is due to its low magnetization. Hence, it suggested the MgO/CoFeB/Ta thin film can be a good candidate to easily tune the PMA strength. Further study shows that the PMA strength can be simply tuned by MgO thickness due to the interface structure between CoFeB and MgO layers [24]. From the definition of S_{AHE}, there are two kinds of methods to achieve the sensitivity of AHE: one is to reduce the H_S and another is to enhance the R_{AH}. Thus, a large linear R_{xy} - H loop can be obtained, when $K_{eff} < 0$ but is very close to 0. And then, a high sensitivity of AHE can be obtained in the perpendicular CoFeB thin film.

4. Summary
In summary, the PMA properties of CoFeB sandwiched between MgO and tantalum layers have been systematically investigated using AHE and PNR. A large PMA in the CoFeB above MgO layer is due to its low magnetization. Furthermore, we demonstrate a giant linear anomalous Hall effect in the perpendicular CoFeB thin film. Our results open a new avenue to detect weak magnetic field by using highly sensitive AHE of perpendicular CoFeB

References
[1] Zhu T 2014 *Chin. Phys. B* **23** 047504
[2] Zhu T, Chen P, Zhang Q H, Yu R C and Liu B G 2014 *Appl. Phys. Lett.* **104** 202404
[3] Freitas P P, Ferreira R, Cardoso S and Cardoso F 2007 *J. Phys.: Condens. Matter* **19** 165221
[4] Manandhar P *et al.* 2009 *Nanotechnology* **20** 355501
[5] Gilbertson A M *et al.* 2011 *Appl. Phys. Lett.* **98** 062106
[6] Teixeira J M *et al.* 2012 *J. Appl. Phys.* **111** 053930
[7] Gerber A 2007 *J. Magn. Magn. Mater.* **310** 2749
[8] Zhu Y and Cai J W 2007 *Appl. Phys. Lett.* **90** 012104
[9] Lu Y M, Cai J W, Pan H Y and Sun L 2012 *Appl. Phys. Lett.* **100** 022404
[10] Ikeda S *et al.* 2010 *Nature Mater.* **9** 721
[11] Worledge D C *et al.* 2011 *Appl. Phys. Lett.* **98** 022501
[12] Zhu T, Yang Y, Yu R C, Ambaye H, Lauter V and Xiao J Q 2012 *Appl. Phys. Lett.* **100** 202406
[13] Lauter V, Ambaye H, Goyette R, Lee W H and Parizzi A 2009 *Physica B* **404** 2543
[14] Bland J A C, Bateson R D, Heinrich B, Celinski Z and Lauter H J 1992 *J. Magn. Magn. Mater.* **104–107** 1909
[15] Feldman B and Dunham S T 2009 *Appl. Phys.Lett.* **95** 222101
[16] Jen S U *et al.* 2006 *J. Appl. Phys.* **99** 053701
[17] Moritz J, Rodmacq B, Auffret S and Dieny B 2008 *J. Phys. D: Appl. Phys.* **41** 135001
[18] Rauch H and Petraschek D 1978 Neutron Diffraction *Topics in Current Physics* vol 6 ed H. Dachs, (Springer, New York)
[19] Felcher G P 1993 *Physica B* **192** 137
[20] Yamanouchi M *et al.* 2011 *J. Appl. Phys.* **109** 07C712
[21] Jang S Y, Lim S H and Lee S R 2010 *J. Appl. Phys.* **107** 09C707
[22] Meng H, Lum W H, Sbiaa R, Lua S Y H and Tan H K 2011 *J. Appl. Phys.* **110** 033904
[23] Sears V F 1992 *Neutron News* **3** 26
[24] Zhu T, Chen P, Zhang Q H, Yu R C and Liu B G *arXiv*:1405.2551

Acknowledgement
This work has been supported by the National Basic Research Program of China (2012CB933102) and National Science Foundation of China (Grants 11574375 and 11174354). Research at Oak Ridge National Laboratory's Spallation Neutron Source was sponsored by the Scientific User Facilities Division, Office of Basic Energy Sciences, US Department of Energy. The author thanks Dr. V. Lauter and Dr. H. Ambaye from ORNL for her help with the PNR measurement and analysis.

Enhanced Magnetism in Field-Cooled $[Ni_{80}Fe_{20}/Mn]_3$ Multilayers Studied Using Polarized Neutron Reflectometry

W. Uilhoorn

School of Physics, The University of New South Wales, Sydney, NSW 2052, Australia
Bragg Institute, Australian Nuclear Science and Technology Organization, Lucas Heights, NSW 2234, Australia

S. J. Callori [1]

School of Physics, The University of New South Wales, Sydney, NSW 2052, Australia
Bragg Institute, Australian Nuclear Science and Technology Organization, Lucas Heights, NSW 2234, Australia

E-mail: sara.callori@csusb.edu

D.L. Cortie

Bragg Institute, Australian Nuclear Science and Technology Organization, Lucas Heights, NSW 2234, Australia
The Institute for Superconducting and Electronic Materials, The University of Wollongong, Wollongong, NSW 2522, Australia

H.-C. Su

National Synchrotron Radiation Research Center, Hsinchu 30076, Taiwan

Y. Khaydukov

Max-Planck Institute for Solid State Research, Stuttgart 50769, Germany
Max-Planck Society Outstation, FRM-II, Garching 85748, Germany

K.-W. Lin

Department of Materials Science and Engineering, National Chung Hsing University, Taichung 402, Taiwan

F. Klose

The Bragg Institute, Australian Nuclear Science and Technology Organization, Lucas Heights, NSW 2234, Australia
Department of Physics and Materials Science, City University of Hong Kong, Hong Kong SAR, China

[1] Now at the Department of Physics, California State University San Bernardino, San Bernardino, California, USA

Abstract. Here, the interfacial magnetic coupling in an exchange biased $[Ni_{80}Fe_{20}/Mn]_3$ multilayer system has been studied using polarized neutron reflectometry. Previous results on this system indicate the importance of the coupling between the Fe-Mn and Ni-Mn orbitals at the layer interfaces. Magnetic depth profiles of the multilayer were measured at low temperatures under field-cooled and zero-field-cooled conditions. While no definitive interfacial state was found, a magnetic moment enhancement of roughly 20-30% in the applied field direction was observed throughout the bulk of the NiFe layers in the field-cooled state as compared to the zero-field-cooled measurements. The origin of this enhancement also likely stems from Fe-Mn and Ni-Mn orbital coupling, but due to the interfacial roughnesses of the sample, the areas where this coupling plays an important role is no longer confined to the interface.

Layered magnetic structures consisting of neighboring ferromagnets (FM) and antiferromagnets (AFM) have played a key role in almost all facets of electronic technology since the discovery of giant magnetoresistance. A hallmark property of these systems is the exchange coupling between the two types of magnetic materials, which can lead to enhanced coercivity and exchange bias, namely as a shift in the hysteresis loop [1]. The use of this behavior to essentially "set" the magnetization direction of a FM layer is one of the main features of the read and write components of magnetic memory devices. The coupling between FM and AFM layers is especially dependent on the interfacial properties of the system, where roughness can be used to tailor exchange coupling [2, 3, 4]. Recent work has also found that spin-orbit coupling can play a role in augmenting exchange bias by enhancing orbital moments in the FM layers near the interface [5].

Here, we report on a $Ni_{80}Fe_{20}$ (NiFe)/Mn (FM/AFM) multilayer system studied using polarized neutron reflectometry (PNR). Previous results on similar samples have demonstrated the onset of exchange bias and enhanced coercivity at low temperatures and x-ray magnetic circular dichroism (XMCD) measurements showed evidence of an increased orbital moment under field-cooling conditions for both the Fe and Ni atoms as a result of overlapping d orbitals between Fe-Mn and Ni-Mn [6]. Such interactions could lead to an enhanced magnetic moment at the NiFe/Mn interfaces. The quality of the interfaces within the system also plays a role in determining its properties. Ion-beam bombardment was used to tailor the exchange bias effects in NiFe/Mn bilayers and was demonstrated to alter the Mn spin structure near the interface [7]. When extended to multiple repetitions and compared to composite [NiFe-Mn] layers, the presence of distinct interfaces also dictated the dominant magnetic coupling observed [8].

Due to the importance of the interface in the system presented here, it is worthwhile to correlate the structure of the multilayer and its magnetism. PNR offers a way to obtain a depth-resolved profile of both the nuclear and in-plane magnetic structure of thin films and layered systems and can thus be used to connect structural features such as disorder or interfaces with the magnetic moment at a given depth. Therefore PNR is an apt choice to study potential interfacial magnetic enhancement in NiFe/Mn multilayers.

1. Experiment

A [NiFe ($\sim$ 3.5nm) / Mn ($\sim$ 3.5nm)]$_3$ sample was deposited on an oxidized Si substrate using a dual ion-beam deposition technique, with conditions described in Ref. 6. A 6 nm Al capping layer was deposited on top of the multilayer stack. Due to the relatively low Neél temperature of the AFM Mn layers ($\sim$100 K) [6], no magnetic field was applied during deposition. Fig. 1 shows a schematic of the multilayer sample.

PNR was performed on the $[NiFe/Mn]_3$ multilayer in order to probe the magnetic moments in the FM NiFe layers and search for any interfacial effects. PNR uses small angle scattering in a reflection geometry to probe both the magnetic and nuclear depth profiles across a thin film or multilayer system. PNR provides information about the scattering length density (SLD)

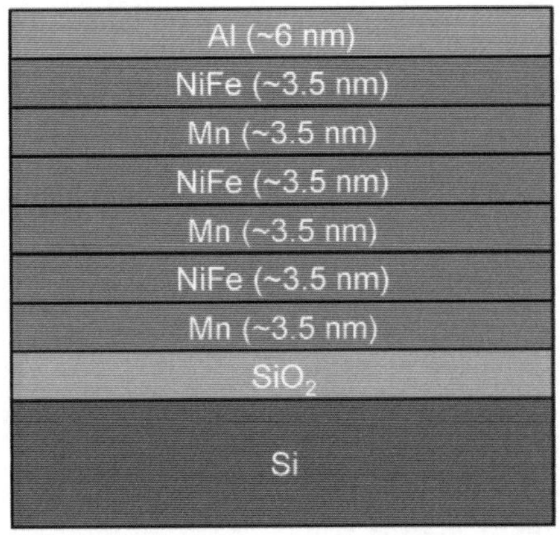

Figure 1. Schematic of the $[NiFe/Mn]_3$ multilayer sample investigated using PNR.

in the direction the of the scattering vector, Q_z, which is perpendicular to the sample surface, and is given as $Q_z = 4\pi\frac{\sin\theta}{\lambda}$, where λ is the neutron wavelength and θ the angle of the incident beam. The SLD contains both nuclear (e.g. atomic density and chemical makeup) and magnetic information with $SLD_{total} = SLD_{nuclear} \pm SLD_{magnetic}$ depending on the neutron polarization (R+ or R−). From measuring both R+ and R− spin channels, where + or − gives the spin state of the incoming neutrons, the difference in SLD yields the magnetic SLD, which can be used to obtain a profile of the in-plane magnetic moment parallel to the neutron spin and applied field across the system. PNR measurements were made at the NREX reflectometer at the FRM-II research reactor in Germany. The program SIMULREFLEC was used to model the data and obtain SLD profiles as well as the magnetic moments of each layer [9].

2. Results

Magnetometry results on this system can be found in Ref. [6] and Ref. [8]. Both an increased coercivity and exchange bias behavior in Magnetization-Field hysteresis loops at low temperatures are clear evidence of exchange coupling [6]. Magnetization vs. Temperature studies show a large discrepancy between the field-cooled and zero-field cooled magnetic states at low temperatures [6]. Below roughly 50 K, there was a significant increase in the overall magnetization of the multilayer as measured by magnetometry for the field-cooled state.

In order to determine the origins of this behavior, PNR was used to obtain magnetic depth profiles of the system under field-cooled and zero-field-cooled conditions. First, the sample was field-cooled in 2.5 kOe field to the measurement temperature of 4 K, where a PNR scan was performed. The sample was then heated up to room temperature and re-cooled to 4 K under zero applied field. When this temperature was reached a field of 2.5 kOe was applied and a second PNR profile was obtained. PNR data were collected for the R+ and R− channels, where the magnetization in the direction of the neutron polarization and applied magnetic field can be measured. The resulting reflectivity curves are shown in Fig. 2 for the zero-field-cooled case (a) and field-cooled case (b). The spin asymmetry between the two spin channels (defined as $\frac{R+ - R-}{R+ + R-}$), shown in (c), is proportional to the magnetization within a system.

Both the nuclear and magnetic depth SLD profiles were then modeled as to best fit the PNR data and the SLD profiles can be seen in Fig. 3. The magnetic SLD profiles are shown in Fig. 3 (a) for both the zero-field-cooled and field-cooled states. The magnetism within the sample is constrained to within the NiFe layers for both cases, although the interfaces between the NiFe and Mn layers are less sharp than previous measurements of similar samples, with roughnesses

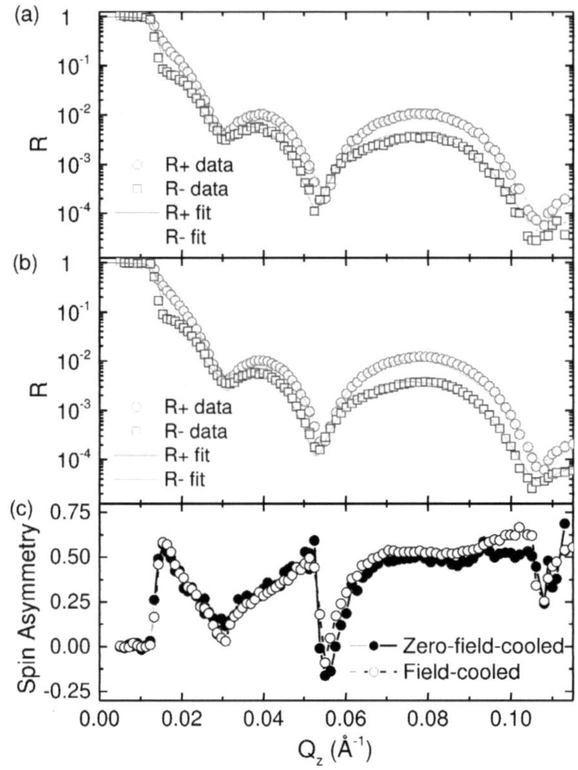

Figure 2. Reflectometry results for the zero-field-cooled (a) and field-cooled (b) states. The circles show the data and the solid lines are the best fit models. (c) Shows the experimental spin asymmetries ($\frac{R+ - R-}{R+ + R-}$), which is proportional to the magnetization of the film, for both conditions. The error bars like within the markers.

Table 1. The components of the magnetic moment parallel to the applied field direction for both cooling scenarios.

Cooling Condition	Bottom NiFe	Middle NiFe	Top NiFe
Zero-field-cooled	0.9 $\mu_B/f.u.$	1 $\mu_B/f.u.$	0.8 $\mu_B/f.u.$
Field-cooled	1.1 $\mu_B/f.u.$	1.35 $\mu_B/f.u.$	0.9 $\mu_B/f.u.$

of approximately 10 to 20 Å, as opposed to 5 Å in Ref. 6. This may also result from the increased contrast between NiFe and Mn due to the negative SLD on Mn, as compared to the positive x-ray SLD's of both materials, which was used to ascertain interfacial roughness values in [6]. A striking increase is noticeable between the magnetic SLD values for the field-cooled case as compared to the zero-field-cooled values. Table 1 shows the calculated moment for each NiFe layer, where the bulk moment is 0.9 - 1 $\mu_B/f.u.$ [7].

3. Discussion

The best fit SLDs presented in Fig. 3 represent a model where it was not necessary to introduce an interfacial layer that was magnetically distinct from the bulk of the NiFe layers. The addition of an interfacial layer with either an enhanced or reduced magnetic moment did not significantly

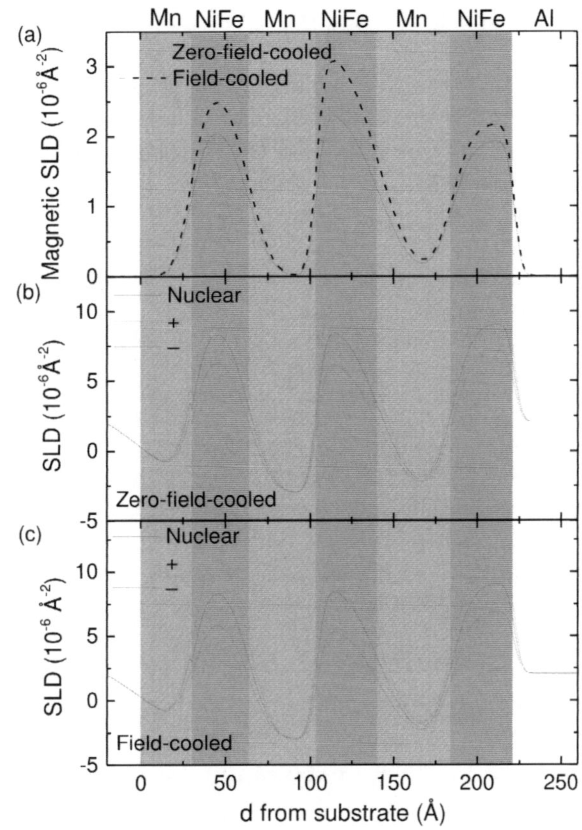

Figure 3. SLD profiles for the fits that best fit the zero-field-cooled and field-cooled PNR measurements. The magnetic SLD profiles are shown for both states in (a). The SLD profiles Nuclear scattering and Nuclear ± Magnetic (+ and −) scattering are shown in (b) for the zero-field-cooled case and (c) for the field-cooled case.

improve or worsen the fit to the field-cooled data, so the presence of such a layer cannot be conclusively ruled out. The first scenario, of an enhanced interfacial moment, is what would likely be expected as a result of an orbital overlap across the interface observed in the field-cooled case, as was reported in Ref. 6. The second case, a reduced interfacial moment, may be indicative of a perpendicular anisotropy resulting from spin-orbit coupling at the interface [11], which has been observed in similar NiFe/Ru multilayers [12].

Although the fits are inconclusive with regards to an interfacial layer, it is striking that there is a relatively large enhancement of the NiFe moments throughout the bulk of the layers for the field-cooled measurements. When comparing the NiFe layers to previous data, it is also apparent that we are clearly looking at an *enhancement* for the field-cooled case rather than a reduction for the zero-field-cooled case. With the exception of the top NiFe layer, the bottom and middle NiFe layers have magnetic moments equal to that of the bulk value in the zero-field-cooled measurements. Additionally, magnetometry measurements on a similar sample shown in Ref. 6 show a coercive field of less than 500 Oe in the field-cooling direction, so it is evident that the PNR measurements are made in a saturated state.

The observation of the enhancement throughout the NiFe layers is likely a result of the increased interfacial roughness in this particular sample. Higher roughness would lead to a widened interface region with areas of potentially intermixed NiFe and Mn, causing a larger amount of Ni and Fe atoms to neighbor Mn sites. As can be seen from the gradual change of SLD's from Mn to NiFe in Fig. 3, the interfacial regions penetrate into a large fraction of the NiFe layers, which do not reach their maximum SLD values until almost the center of each layer, so an enhanced orbital moment may be expected throughout nearly the entirety of each NiFe layer rather than in only a layer near the interface. This would also explain why the degree of magnetization enhancement is much less in the top NiFe layer. This layer is only bordered by a single Mn layer and a larger fraction of this NiFe layer is near the bulk NiFe SLD value, indicating a smaller amount of NiFe and Mn mixing.

When comparing this conclusion to the results from Ref. 6, where the magnetization nearly doubled between the zero-field-cooled and field-cooled magnetometry measurements, here a magnetization enhancement of only roughly 20-30 % was seen in the bottom and middle NiFe layers. However, the magnetometry was performed under only a 100 Oe field, as compared to the 2.5 kOe measurement field here, so it is likely that previous results may not have been made in a completely saturated state. The increase in the NiFe moment observed by PNR does agree more with that observed in the XMCD measurements. Using XMCD an orbital enhancement of 15-20% was observed at 20 K. It is likely that here we are either seeing a further increase due to a lower measurement temperature (4 K) or that we are observing the same degree of orbital enhancement [6] with a slight additional increase of the spin moment.

4. Conclusions

Here we have presented a PNR study of a $[NiFe/Mn]_3$ multilayer system. PNR was used to obtain magnetic depth profiles of the system in both field-cooled and zero-field-cooled states. There was no evidence of an interfacial layer with either an increased or decreased magnetic moment in the field-cooled condition. However, an enhanced moment was observed in the bulk of the NiFe layers in the field-cooled state when compared to the field-cooled layer magnetizations and the bulk magnetic moment. Much as in previously reported results, the origins of this increased moment are likely to be from Fe-Mn and Ni-Mn orbital coupling. The larger interfacial roughness observed here result in more intermixing of the NiFe and Mn atoms, allowing the enhanced state to persist throughout the NiFe layer rather than to remain confined to the interface.

References

[1] Nogués J and Schuller I K 1999 *J. Magn. Magn. Mater.* **192** 203
[2] Kuch W, Chelaru L I, Offi F, Wang J, Kotsugi M, and Kirschner J 2006 *Nature Mater.* **5** 128
[3] Spray J and Nowak U 2006 *J. Phys. D: Appl. Phys.* **39** 4536
[4] Lin K-W, Mirza M, Shueh C, Huang H-R, Hsu H-F, van Lierop J 2012 *Appl. Phys. Lett.* **100** 122409
[5] Schmitz D, Schierle E, Darowski N, Maletta H, Weschke E, and Gruyters M 2010 *Phys. Rev. B* **81** 224422
[6] Su H-C, Huang M-J , Lin H-J, Lee C-H, Chen C-T, Liu C-H, Hsu H-F, Lin K-W, and van Lierop J 2013 *Phys. Rev. B* **87** 014402
[7] Lin K-W, Chen T-J, Guo J-Y, Ouyang H, Wei D-H, and van Lierop J 2009 *J. Appl. Phys* **105**, 07D710
[8] Liu C-H, Shueh C, Lan T-C, Lin K-W, Chen W-C, Wu T-H, Desautels R D, and van Lierop J 2013 *J. Korean Phys. Soc.* **62**, 1958
[9] SimulReflec, Copyright (C) Lab. Lèon Brillouin CEA/CNRS UMR12. Free software available at http://www-llb.cea.fr/prism/programs/simulreflec/simulreflec.html.
[10] Cortie D L, Ting Y-W , Chen P-S, Tan X, Lin K-W, and Klose F 2014 *J. Appl. Phys* **115**, 073901
[11] Bruno P 1989 *Phys. Rev. B* **39**, 865
[12] Lin M-Z, Su H-C, Lee C-H, and Huang J C A 2007 *Surf. Sci.* **601**, 5707

Quantum entanglement in manganese(II) hexakisimidazole nitrate: on electronic structure imaging - A polarized neutron diffraction and DFT study

Warren A. Wallace[1]

1. Laboratory for Neutron Scattering and Imaging, Paul Scherrer Institut, Villigen PSI, CH-5232, Switzerland.
Email: warrenwallace@gmail.com

Abstract. Quantum entanglement has been visualized for the first time, in view of the spin density distribution and electronic structure for manganese in manganese(II)hexakisimidazole nitrate. Using polarized neutron diffraction and density functional theory modelling we have found for the complex, which crystallizes in the $R\bar{3}$ spacegroup, a = b = 12.4898(3) Å, c = 14.5526(4) Å, $\alpha = \gamma = 90^0$, $\beta = 120^0$, Z = 3, that spatially antisymmetric and spatially symmetric shaped regions of negative spin density, in the spin density map for manganese, are a result of quantum entanglement of the high spin d^5 configuration due to dative imidazole-manganese π- donation and σ-bonding interactions respectively. We have found leakage of the entangled states for manganese observed as regions of positive spin density with spherical (3.758(2) μ_B) and non-spherical (1.242(3) μ_B) contributions. Our results, which are supportive of Einstein's theory of general relativity, provide evidence for the existence of a black hole spin density distribution at the origin of an electronic structure and also address the paradoxical views of entanglement and quantum mechanics. We have also found the complex, which is an insulator, to be suitable for spintronic studies.

1. Introduction

In condensed matter, the state and structure of the electron is assessed through interactions with its environment. Quantum mechanics[1] suggests, for instance, for the entangled interactions between spin systems $S_A = |i\rangle_A$ and $S_B = |j\rangle_B$ in Hilbert space:

$$|\psi\rangle_{AB} = \sum_{i,j} c_{i,j} |i\rangle_A \otimes |j\rangle_B \quad (1a)$$

that the final state of the system, $|\psi\rangle_{AB}$, disentangles when either $|i\rangle_A$ or $|j\rangle_B$ is measured providing:

$$c_{ij} = c_i^A c_j^B \quad (1b)$$

where $c_{ij.}$ is the coefficient matrix in the subspace ij.

Otherwise, the final state of the system $|\psi\rangle_{AB}$ remains entangled (correlated) providing:

$$c_{ij} \neq c_i^A c_j^B \quad (1c)$$

The quantum mechanical interaction of a spin-½ electron with an applied field $\mathbf{B}_z$ was first demonstrated in the Stern-Gerlach[2] experiment (figure 1, below).

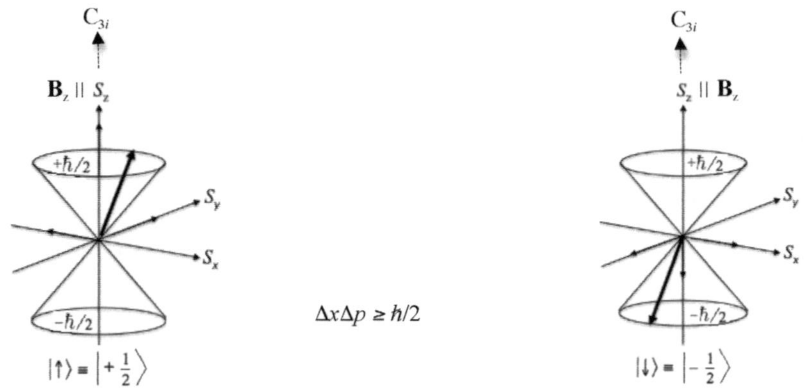

Figure 1: Precession cones associated with a spin-½ electron in applied field $\mathbf{B}_z$

In figure 1, the doubly degenerate spatially antisymmetric precession cones about $\mathbf{B}_z$, for the quantum entangled spin states $|\pm½\rangle$ of the spin-½ electron, confer an intrinsic C_{3i} point group. The C_{3i} point group breaks spherical (O_h) rotational symmetry about $\mathbf{B}_z$ by lifting the $t_{2g}(O_h)$ level and lowering the $e_g(O_h)$ level. The $t_{2g}(O_h)$ is transformed according to the spatially antisymmetric doubly degenerate E_g irreducible representation and the $e_g(O_h)$ transformed according to the spatially symmetric singly degenerate A_g irreducible representation, of the C_{3i} point group. The doubly degenerate spatially antisymmetric precession cones about $\mathbf{B}_z$, for the quantum entangled spin states $|\pm½\rangle$, suggest spin delocalization in the transformed $t_{2g}(O_h)$ associated with the E_g irreducible representation. By contrast, since there is no spatially symmetric precession about $\mathbf{B}_z$ a vacuum (spherical solution) is suggested for the transformed $e_g(O_h)$ associated with the singly degenerate A_g irreducible representation. We propose these qualitative symmetry assessments for the spin-½ electron in an applied field $\mathbf{B}_z$ reflect the intrinsic point group of the electron with respect to its magnetic moment.

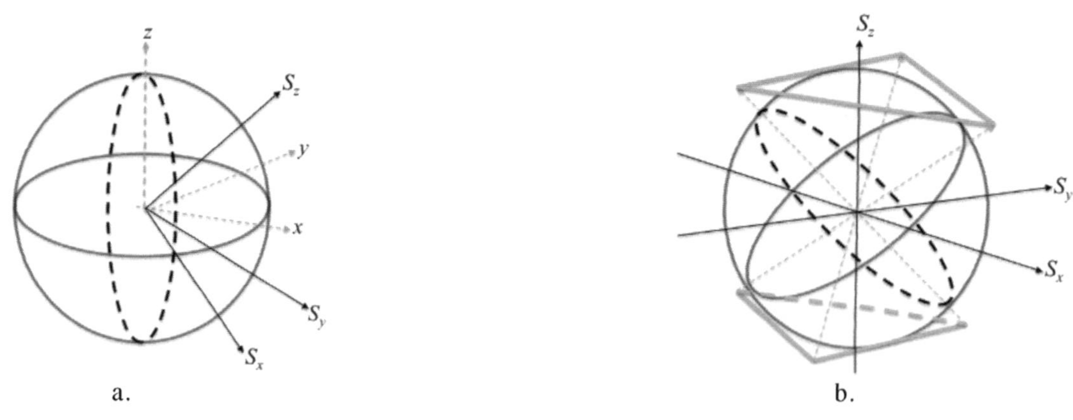

Figure 2: proposed schematic decomposing classical spherical symmetry for a spin-½ electron into irreducible states; a doubly degenerate spatially antisymmetric state (green-colored triangles), and a singly degenerate spatially symmetric state (black-lined sphere). Note: in b.) S_z is arbitrarily chosen as the spin angular momentum quantization vector.

Given figure 2b, we suggest that the spin angular momentum operator S for a spin-½ electron with an intrinsic electric field $\boldsymbol{E}$ be described as:

$$\mathbf{S} = \frac{\epsilon_0}{2i\omega} \sum_{i=x,y,z} \int (E^{i*}(\mathrm{r} \times \nabla) E^{i}) d^3\mathrm{r} \tag{2}$$

where ϵ_0 = the permittivity of vacuum in free space and i = cartesian components. Also, from figure 2b and equation (2), the electron's spin angular momentum arises from intrinsic spin precession (spin delocalization) in three-dimensional space.

We propose, according to figure 2b, classical spin entanglement for an electronic structure in Hilbert space, equations (3a - b) below. We describe a spatially antisymmetric classical entangled ferromagnetic spin exchange state, in the two-particle state problem:

$$|\Psi(t)\rangle = \frac{n}{\sqrt{2}}(|\downarrow\rangle_1 \otimes |\uparrow\rangle_2 - |\uparrow\rangle_1 \otimes |\downarrow\rangle_2) \tag{3a}$$

which is fermion-like, where n = 3 for three-fold orbital symmetry. We also describe a spatially symmetric classical entangled antiferromagnetic spin exchange state, which is boson-like:

$$|\Psi(t)\rangle = \frac{n}{\sqrt{2}}(|\downarrow\rangle_1 \otimes |\uparrow\rangle_2 + |\uparrow\rangle_1 \otimes |\downarrow\rangle_2) \tag{3b}$$

where, n = 1 for spherical orbital symmetry. The simultaneous occurrence of both fermion-like and boson-like spatial states in equations (3a and 3b), for our proposed description of an electronic structure in Hilbert space, is a likely consequence of an anyon-like[3] intermediate during the two–particle spin states interchange. Further, since the electron has a rest mass (m_e) and rest energy it intrinsically follows $E_e = m_e c^2$ in view of Einstein's mass-energy[4] equivalence. In this respect, the boson-like state in equation (3b) and the fermion-like state in equation (3a) define the mass-like and light-like nature, respectively, of our proposed electronic structure in Hilbert space. Our conclusions, in this respect, are akin to an observer's (mass-like) view of "past" and "future" (light-like) cones in Minkowski[5] general relativistic space-time (figure 3, below).

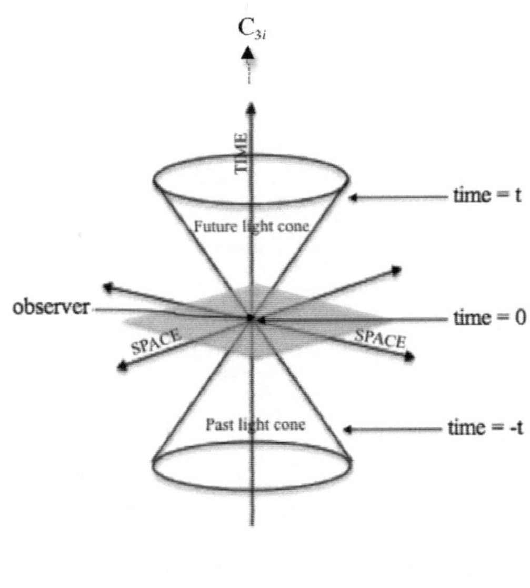

Figure 3: Minkowski[5] general relativistic space-time for an observer's view of "past" and "future" light cones.

The paradox of quantum formalism implies that a measurement in Planck's time, for example through the application of an applied field, on a member of a two-particle classical entangled spin exchange state collapses the classical system before any space–like information between entangled members can be transmitted. In this work we will consider Einstein's possible solution[6] to this paradox through our proposed view that hidden information, such as spatial symmetry (which confer spin distribution symmetry) and energy–momentum (which confer spin delocalization), pertaining to the classical entangled spin exchange state in Hilbert space, is carried by each classical entangled member and no information is transmitted between members during a quantum measurement. In this respect, we suggest the system remains entangled in the quantum world, albeit in a quantum entangled spin exchange state in Hilbert space (figure 4 below), during a quantum measurement on a member of the classical entangled world. Also, spatial and energy-momentum information leaked in the quantum entangled world will be representative of the hidden information of the classical entangled world. Furthermore, since we have proposed the coexistence of spatially antisymmetric and spatially symmetric classical entangled spin exchange states (equations (3a – b)) to be in view of the light-like and mass-like nature respectively of our proposed electronic structure in Hilbert space then their simultaneous assessment during measurement will be expected.

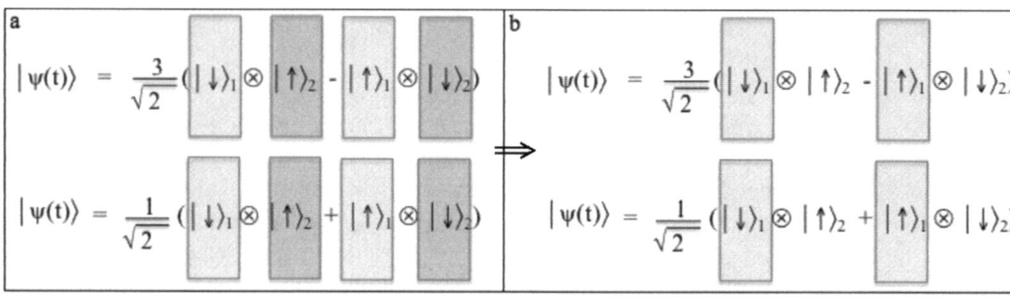

Figure 4 a. Proposed example of members of the classical entangled world (color-coded green and purple) in Hilbert space; and b. proposed example of a quantum entangled world in Hilbert space at the onset ($\Rightarrow$) of a measurement on a member of the classical world.

Finally, we assume '*action at a distance*'[6] (that is, entanglement which is implicit of electron correlation in our discussion) to be relative. We note the premise of Einstein's theory of general relativity[7] that gravity is an intrinsic geometric property of space–time and that the curvature of space–time is related to the energy, momentum of matter present is in support of our proposed electronic structure in Hilbert space (equations (3a – b), figure 2b). For instance, since we have suggested no spin angular momentum is associated with the classical spatially symmetric (boson-like) entangled state (equations (2, 3b), figure 2) then Minkowski[5] space–time (figure 3) for the associated spin density distribution, during a quantum measurement, is expected to be spherical and localized (at $t = 0$, the origin in figure 3). This spherical solution, referred to as the vacuum solution to Einstein's field equations (EFE)[8-9] since the energy–momentum tensor in EFE[8-9] is zero, is defined by the Schwarzschild metric[10-11]. Further, a black hole[12], an object with a radius smaller than the Schwarzschild radius[10-11] and having a predicted singularity[13-14], is also expected to be found. (This suggests, within the context of the associated spin density distribution, a 'black hole spin density distribution' at the origin). By contrast if the energy–momentum tensor in EFE[8-9] is not zero Minkowski[5] space–time (figure 3) is spatially antisymmetric (at $t = +t$ and $t = -t$ in figure 3). In this respect, we expect during measurement that the spin density distribution associated with our proposed classical spatially antisymmetric (fermion-like) entangled state, which has intrinsic orbital angular momentum (equations (2, 3a), figure 2b), to be delocalized and spatially antisymmetric having C_{3i} symmetry.

Journal of Physics: Conference Series **711** (2016) 012006 doi:10.1088/1742-6596/711/1/012006

2. Experimental overview

To explore experimentally and theoretically the aforementioned considerations, we investigate the spin density distribution and electronic structure for the interactions of manganese in the titled complex (figure 5 below) using polarised neutron diffraction (PND)[15] and density functional theory modeling.

a.

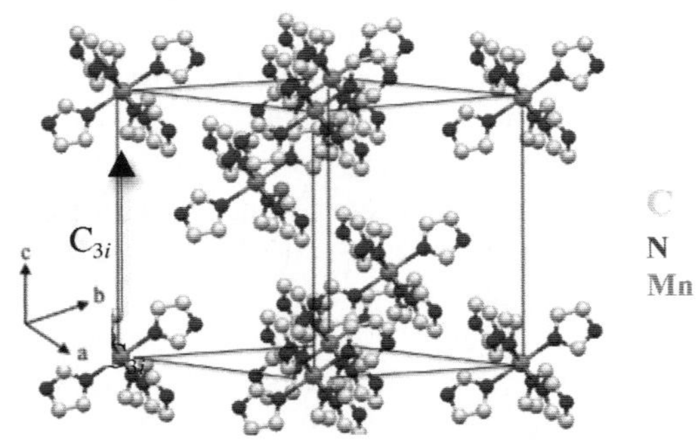

b.

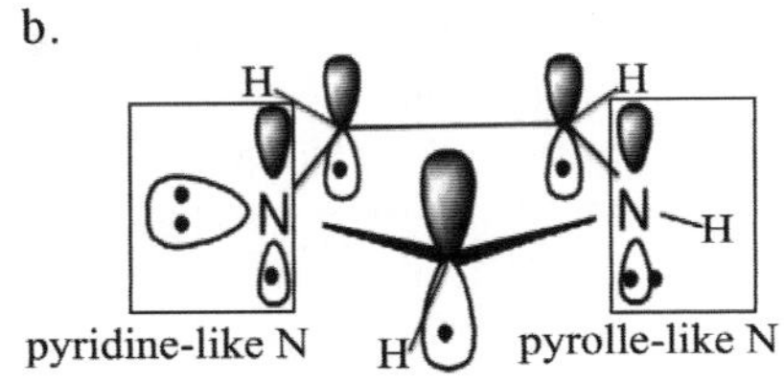

Figure 5: a. Trigonal unit cell for the titled complex. For clarity, hydrogen atoms and nitrate counter ions have been omitted; b. Schematic for imidazole, showing the pyridine-like nitrogen which coordinates to manganese in the titled complex.

2.1 *Classical spin exchange entanglement, in Hilbert space, for the interactions of manganese and imidazole in the titled complex.*

We propose[16] that if the high spin d^5-configuration for Mn(II) has the α-spin direction, or the β-spin direction, an occupied Φ_{3d} α-spin orbital, or an occupied Φ_{3d} β-spin orbital, is defined. The Lewis acidity of Mn(II) in this context, with respect to a coordinating imidazole moiety, is viewed as an unoccupied Φ_{3d} spin orbital on Mn(II) in the presence of an occupied spin orbital associated with a coordinating pyridine-like nitrogen of imidazole (Figure 5b), where imidazole acts as a Lewis base. In this respect, while considering the valence bond (also Heitler-London model), localized bonding, independent bond approaches,[16-21] we suggest classical spin entanglement for an electronic structure in Hilbert space (equations (3a - b), figure 2b) associated with the interactions of manganese and imidazole in the titled complex. Within the context of spin states interaction, we suggest:

- *Dative imidazole-manganese σ -bonding interaction*, involving the spin states (arbitrarily assigned $\alpha(\uparrow)$ and β $(\downarrow)$) for the pair of electrons in the available sp^2-hybridized orbital of the pyridine-like nitrogen of imidazole interacting with an unoccupied Φ_{3d} spin orbital associated with the e_g set of Mn(II). Additionally, we propose[16] that dative imidazole-manganese σ-bonding interactions constitute antiferromagnetic spin exchange interaction, in the two particle problem, such that a classical spatially symmetric entangled exchange state in Hilbert space is defined as shown in equation (3b).

- *Dative imidazole-manganese π-donation interaction*, involving the unpaired electron in the p_z orbital of the pyridine-like nitrogen of the coordinating imidazole π-donating unto an unoccupied Φ_{3d} spin orbital associated with the t_{2g} set of Mn(II). This constitutes ferromagnetic spin exchange interaction, in the two particle state problem, such that a classical spatially antisymmetric entangled exchange state in Hilbert space is conferred equation (3a)).

Journal of Physics: Conference Series **711** (2016) 012006 doi:10.1088/1742-6596/711/1/012006

2.2 *Leakage of hidden information (spatial symmetry, energy-momentum) during a PND measurement of the titled complex*:

- *For dative imidazole-manganese σ -bonding interaction* we suggest the leaked spatial information during measurement, associated with our proposed classical spatially symmetric entangled state in Hilbert space (equation (3b)), be described as the spherical symmetric $^6A_{1g}$ (t_{2g}^3 e_g^2) high spin crystal field state for manganese associated with the σ* orbital on manganese. Also, since no orbital angular momentum is associated with dative imidazole-manganese σ-bonding interactions[16], as the σ-bond quantization axis taken as the z-direction is projected towards the manganese nucleus, no spin delocalization (that is, zero energy-momentum) is associated with the $^6A_{1g}(t_{2g}^3\ e_g^2)$ state. Further, in accordance with the theory of general relativity[7], a zero value for the energy-momentum tensor in EFE[8-9] imparts a vacuum solution to EFE[8-9]. This also confers an implicit black hole[12] having a radius smaller than the Schwarzschild radius[9-10]. We therefore expect an imaged black hole spin density distribution for our proposed $^6A_{1g}(t_{2g}^3\ e_g^2)$ state for manganese during measurement. Analytically we define the measured magnetic moment $\mu\sigma^*(Mn)$, associated with the hole in the σ* orbital on manganese, as:

$$\mu\sigma^*(\mathrm{Mn}) = (1 - k_{\mathrm{Mn}\sigma}n_1)g\mu_B S \qquad (4)$$

with $S = 5/2$, $n_1 = 1$ for the spherical $^6A_{1g}(t_{2g}^3\ e_g^2)$ crystal field state, $k_{\mathrm{Mn},\sigma}$ is the admixture coefficient on manganese (assessed from DFT modeling) due to dative imidazole-manganese σ-bonding interaction, $g = 2$ for the electron g–factor, μ_B is the Bohr magneton.

- *For dative imidazole-manganese π-donation interaction* we suggest the leaked spatial information during measurement, associated with our proposed classical spatially antisymmetric entangled state in Hilbert space (equation (3a)), be described as the three-fold spatially antisymmetric $^2T_{1g}(t_{2g}^5\ e_g^0)$ low spin crystal field state for manganese associated with the π* orbital on manganese. Also, since orbital angular momentum is intrinsic to dative imidazole-manganese π-donation interactions[16], as the π-donation quantization axis taken as the z-direction is projected away from the manganese nucleus, spin delocalization (energy-momentum) is associated with the $^2T_{1g}(t_{2g}^5\ e_g^0)$ state. We define the measured magnetic moment $\mu\pi^*$ (Mn), associated with the π* orbital on manganese, as:

$$\mu_{\pi^*}(\mathrm{Mn}) = (1 - (n_2\eta_{\mathrm{Mn}} + n_2 k_{\mathrm{Mn},\pi}))g\,\mu_B\,S \qquad (5)$$

with $S = 5/2$, $n_2 = 3$ for the three-fold $^2T_{1g}$ crystal field state, $k_{\mathrm{Mn},\pi}$ is the admixture coefficient on manganese (assessed from DFT modeling) due to dative imidazole-manganese π-donation interactions, η_M is the spin delocalization for manganese.

Since π-donation by the coordinated imidazoles unto the t_{2g} set of manganese breaks the aromaticity of the imidazoles then spin delocalization $\eta_{\mathrm{Imz(Mn)}}$ for manganese (associated with the $^2T_{1g}$ state) is expected to be observed in the π* orbital of each imidazole. We also expect spin delocalization $\eta_{\mathrm{Imz(Mn)}}$ for manganese to be observed in the σ* orbital of each imidazole. In this respect, we re-define the measured moment for manganese due to dative imidazole-manganese π-donation interaction as:

$$\mu = \mu\pi^*(\mathrm{Mn}) + \mu\pi^*,\sigma^*(\mathrm{Mn(Imz)}) \qquad (6)$$

where, from (6), $\mu\pi^*,\sigma^*(\mathrm{Mn(Imz)})$ is the measured moment for manganese on each coordinate imidazole. We assume $k_{\mathrm{Mn},\pi}$ (in equation (5) = $1/\mu_B(\mu\pi^*(\mathrm{Mn}) - \mu\pi^*,\sigma^*(\mathrm{Mn(Imz)})$.

Comparing equations (4 – 5), we note the value $n_1/n_2 = 1/3$ infers a black hole solution for a three-fold orbital state.

2.3 Given the spin density form factor f_S in q-space:

$$f_S(\mathbf{q}) = \int \rho(\mathbf{r}) e^{i\mathbf{q}\cdot\mathbf{r}} \tag{7a}$$

where, ρ is the spin density and $f_S(\mathbf{q})$ is the fourier transform:

$$f_S(\mathbf{q}) = \int \psi_i^* \psi_j e^{i\mathbf{q}\cdot\mathbf{r}} dr \tag{7b}$$

then, in terms of multipolar expansion[22], the spin density form factor is re-expressed as:

$$f_S(q, \theta, \phi) = \sum \sum i^l M^m{}_l N^m{}_l \langle j_{n,l} \rangle Z^m{}_l (\theta, \phi) \tag{7c}$$

where, $l = 0, 2, 4$ and $-l \leq m \leq l$, $\langle j_{n,l} \rangle$ is the radial integral and, $Z^m{}_l$ (θ, ϕ) are multipole functions (tesseral harmonics):

$$Z^m_l(\theta, \phi) = \begin{cases} Y^m_l \cos\theta \cos m\phi, & m \geq 0 \\ Y^m_l \cos\theta \sin(-m\phi), & m < 0 \end{cases}$$

also, from (7c), N_l^m are normalization functions, M_l^m are multipole population parameters. In assessing our expected observation of spherical and non-spherical contributions to the spin density distribution on manganese in the complex the multipole functions Z_0^0 and $Z_2^0, Z_4^0, Z_4^{3+}, Z_4^{3-}$ will be considered.

3. PND results

Re-construction of the spin density map for manganese (Figure 6a, below), by inverse Fourier[23] of the observed spin density structure factors obtained from the PND experiment on a single crystal of the complex, shows regions of negative and positive spin density. The regions of negative spin density (see figure 6a) are due to spatially antisymmetric quantum entangled spin exchange (*) and spatially symmetric quantum entangled spin exchange (**) states, in Hilbert space, for manganese. We note the quantum entangled spin exchange states for manganese are associated with classical entangled spin exchange states of our proposed electronic structure for the complex in Hilbert space (equations (3a – b), figure 2, section 2.1). In figure 6a, the region of positive spin density at the origin, which was found to have a value $3.758(2)\mu_B$ using Multipolar Least Square Analysis (MPLSQ)[23], is due to leakage of the spherical symmetric $^6A_{1g}(t_{2g}{}^3\ e_g{}^2)$ crystal field state for manganese associated with the spatially symmetric classical entangled state (sections 2.1 – 2.2). As shown in figure 6a, the spin density distribution at the origin constitutes a black hole spin density distribution for manganese. This is because no spin (electron) delocalization (that is, no energy-momentum) is associated with the $^6A_{1g}$ $(t_{2g}{}^3\ e_g{}^2)$ state for manganese. Our observation of a black hole spin density distribution at the origin (see also our discussion at the end of section 1) for manganese in figure 6a is supported by the theory of general relativity[7], which predicts the occurrence of a black hole as an intrinsic consequence of the vacuum solution to EFE[8-9]. This conveys with relevance to dative imidazole-manganese σ-bonding interactions in the titled complex, a collapsing of the mass-like nature of spin density distribution for manganese at the origin, of our proposed electronic structure in Hilbert space (sections 2.1 – 2.2, equations (3a – b), figure 2b). The Schwarzschild radius[10-11] for $S = 5/2$ was calculated to be 0.6765×10^{-54} cm. We have measured a Schwarzschild radius of 0.65 cm based on figure 6a. The factor of 10^{-54} between the calculated and measured Schwarzschild radius is attributed to scaling.

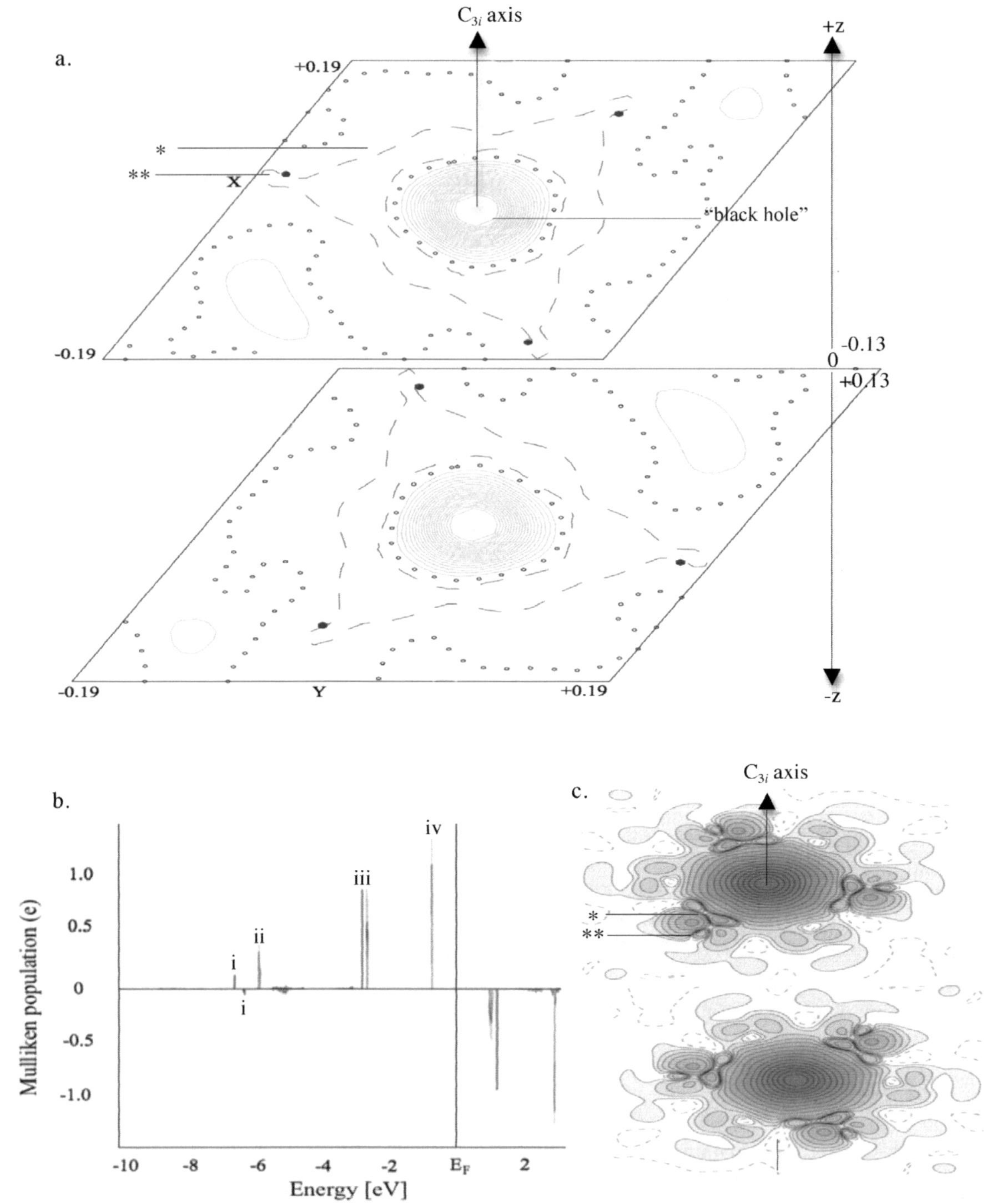

Figure 6: a. Re-construction of the spin density on Mn in the titled complex using the PND data; color code: Mn (yellow) and for clarity prydine-type nitrogen (blue); X parallel to (100),Y parallel to (010), Z parallel to (001); (*) spatially antisymmetric shaped region of negative spin density; (**) spatially symmetric shaped region of negative spin density.
b. DFT modelled Mulliken spin populations on Mn;
c. DFT modelled spin density on Mn
(*Plots of the spin density are 2-dimensional projections parallel to the z-bound projection of the complex*)

In figure 6a the regions of positive spin density projected away (green-colored contours, figure 6a) from the origin, which were assessed to be 1.242(3) μ_B, are a result of spin delocalization (energy-momentum) for manganese. This is due to our proposed leakage of a (non-spherical) three-fold spatially antisymmetric $^2T_{1g}$ (t_{2g}^5 e_g^0) low spin crystal field state for manganese. Spin delocalization of the $^2T_{1g}$ (t_{2g}^5 e_g^0) state for manganese, which is associated with the π* orbital on manganese, occurs also in the π* and σ* orbitals of the coordinated imidazole. In this respect our assessed value of 1.242μ_B for the regions of positive spin density projected away from the origin in figure 6a takes in account equation 6 of section 2.2. Lastly, we conclude the value 1.242μ_B/5μ_B = ¼, which is indicative of the entropy associated with a black hole[24], is a result of orbital angular momentum conferring non-spherical spin density distribution for manganese. This inherently reduces the mass (in Bohr magneton units) of the black hole spin density distribution for manganese at the origin (in figure 6a) from 5μ_B to 3.758μ_B.

4. DFT modeling overview, results

In modeling the spin density for manganese in the titled complex, the Dmol3 package[25-26] was employed. We note the spin density $\rho_\sigma(\mathbf{r})$, given as a sum over all occupation points (k) for the function ($\Phi_{k,}\sigma$):

$$\rho_\sigma(\mathbf{r}) = \sum_k^{occ} |\Phi_{k,\sigma}(\mathbf{r})|^2 \tag{8a}$$

$$\Phi_{k,\sigma}(\mathbf{r}) = \sum c_{j,k,\sigma}\, \phi_j(\mathbf{r}) \tag{8b}$$

also, in spin basis σ, the density is represented:

$$\rho = \rho_\sigma + \rho_{-\sigma} \tag{8c}$$

and, due to variations in density, the total energy E_t minimized is given as:

$$E_t = \sum_k^{occ} \epsilon_k + \sum_\sigma \int \rho_\sigma \left(\epsilon_{xc} - \mu_{xc,\sigma} - \frac{V_e}{2} \right) d\overset{3}{r} + \sum_{\alpha,\beta} \frac{Z_\alpha Z_\beta}{|\mathbf{R}_\alpha - \mathbf{R}_\beta|} \tag{9}$$

where, V_e defines the electrostatic potential, α and β are spin up and spin down states respectively, and μ_{xc} the exchange correlation. A general eigenvalue matrix equation,[25-26] where the matrix elements in the subspace ij are given as:

$$\sum_i h_{i,j}\, c_{j,k} = \epsilon_k \sum_i s_{i,j}\, c_{j,k} \tag{10a}$$

$$h_{i,j} = \int \phi_i(\mathbf{r}) \left\{ \frac{-\hbar}{2m} \Delta + V_s(\mathbf{r}) + \mu_{xc}(\mathbf{r}) \right\} \phi_j(\mathbf{r})\, d\overset{3}{r} \tag{10b}$$

$$s_{i,j} = \int \phi_i(\mathbf{r})\, \phi_j(\mathbf{r})\, d\overset{3}{r} \tag{10c}$$

Dmol3[25-26] uses numeric orbitals $\phi_i(\mathbf{r})$ as the basis set calculated from multipolar functions. The multipole functions, which take into account the point group symmetry of the complex are defined through the expansion:

$$\rho_{\alpha,l,m}(s) = \frac{1}{\sqrt{4\pi}\sqrt{2l+1}} \times \int_{s=|\mathbf{r}-\mathbf{R}|} Y_{l,m}(\mathbf{r}-\mathbf{R}_\alpha)\, \rho_\alpha(\mathbf{r})\, \rho(\mathbf{r})\, d\Omega \tag{11}$$

Journal of Physics: Conference Series **711** (2016) 012006 doi:10.1088/1742-6596/711/1/012006

The results of the DFT modeled spin density on manganese (figure 6c) intrinsically represents a leakage of spin density information consequential of the C_{3i} point group for the complex used to determine the numeric input basis set in equation (11). Comparing the Mulliken spin population plot (figure 6b) with the spin density plot (figure 6c), we note the spatially antisymmetric and spatially symmetric shaped regions of negative spin density in figure 6c are associated with states (i) and (ii) respectively in figure 6b. These are due to off-diagonal matrix elements for equation (10). The diagonal matrix elements for equation (10), attributed to triangular and circular contour regions of positive spin density along the C_{3i} axis in figure 6c, are associated with states (iii) and (iv) in figure 6b. States (iii) and (iv) in figure 6b, which are doubly and singly degenerate respectively, are contributions arising from the transformation of the t_{2g} (O_h) and e_g (O_h) sets for manganese according to the doubly degenerate E_g and singly degenerate A_g irreducible representations respectively of the C_{3i} point group. While states (i) and (ii) in figure 6b, which are doubly and singly degenerate respectively, are contributions arising from the transformations of s-type and p-type orbitals (for the coordinating pyridine-like nitrogen of imidazole) according to the doubly degenerate E_u and singly degenerate A_u irreducible representations respectively of the C_{3i} point group.

In comparing the DFT modeled spin density plot (figure 6c) and the PND spin density plot (figure 6a) the black hole spin density distribution for manganese at the origin in the PND spin density plot is absent in the DFT modeled spin density plot. This difference, we propose, is due to general relativistic[7] considerations of energy–momentum, not taken into account in the DFT modeling which assumes special relativisty[27]. Furthermore, the DFT spin density model assumes complete decoupling of the entangled states shown in figure 7 below.

Figure 7: DFT modeled entangled states for the complex associated with π^* and σ^* orbitals on manganese due to dative imidazole-manganese π-donation interactions and dative imidazole-manganese σ-bonding interactions, respectively.

Given the results of the Mulliken spin population plot in figure 6b the admixture co-efficient for manganese $k_{Mn,\pi}$ (equation (5)) associated with dative imidazole-manganese π-donation interactions (figure 7a) was assessed to be 0.08. While the admixture coefficient for manganese $k_{Mn,\sigma}$ (equation (4)) associated with dative imidazole-manganese σ-bonding interactions (figure 7b) was assessed to be 0.25. Given the general relativistic[7] consideration of energy–momentum, which is in view of spin delocalization η_{Mn} (equation (5)) for manganese associated with dative imidazole-manganese π-donation interactions, the value of the admixture co-efficient for manganese $k_{Mn,\pi}$ was re-assessed to be 0.29 ($\eta_{Mn} + k_{Mn,\pi}$, with $\eta_{Mn} = 0.21$).

5. Electronic structure, suitability of the complex for spintronic studies.

It is well known that quantum entanglement, an implicit consequence of bonding interactions between two spin systems, is associated with the electronic structure arising from antiferromagnetic (spin singlet) exchange and ferromagnetic (spin triplet) exchange in Hilbert space. The antiferromagnetic spin exchange is spatially symmetric and the ferromagnetic spin exchange is spatially antisymmetric. If the exchange interaction is strongly ferromagnetic the spin triplet exchange state is lower in energy than the spin singlet exchange state. Otherwise, if the exchange interaction is strongly antiferromagnetic the spin singlet exchange state is higher in energy than the spin triplet exchange state. Given the results for the titled complex, we suggest the spin triplet exchange state is associated with a non-spherical spin density distribution and the spin singlet exchange state associated with a spherical spin distribution. Using the nuclear structure for the titled complex, we modeled complex stability (figure 8, below) in view of trigonal (non-spherical) stability for the series $[M(Imidazole)_6]^{2+}(NO_3^-)_2$, where M is a 3d-transition metal ion. According to figure 8, complex stability is greater for the d^5 configuration. The titled complex is therefore the best candidate in the series for observing quantum entangled, electronic structure in Hilbert space.

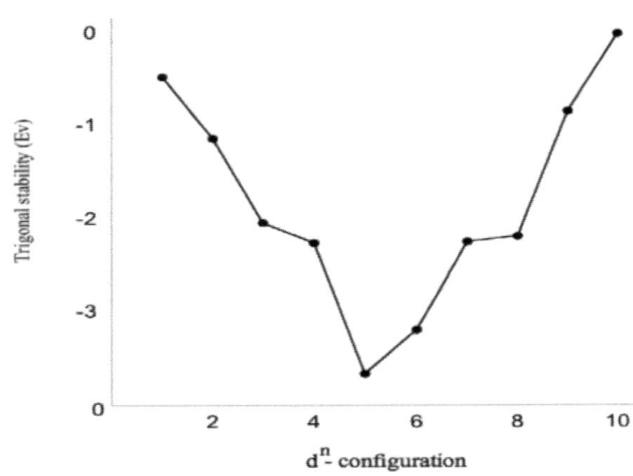

Figure 8: Trigonal stability (Ev) for the series of complexes $[M(Imidazole)_6]^{2+}(NO_3^-)_2$, where M is a 3d-transition metal ion.

The suitability of the titled complex for spintronic studies, such as allowing the flow of spin-polarized currents over its surface, is firstly due to its expected conducting nature (state (iv) (figure 6b), which is spatially singly degenerate, being close to the Fermi level). And, secondly, due to the complex's expected insulating nature (state (iii) (figure 6b), which is spatially doubly degenerate, being further below the Fermi level in the bulk). The black hole spin density distribution for manganese at the origin in the PND spin density map (figure 6a), which we associate with the spherical symmetric $^6A_{1g}(t_{2g}^3\ e_g^2)$ state for manganese, indicate the ability of the titled complex to be spin polarizing at its surface (see also state (iv), figure 6b) to spin polarized currents. While the spin density distribution for manganese projected away from the origin in the PND spin density map (figure 6a), which we attribute to spin delocalization associated with the three-fold spatially antisymmetric $^2T_{1g}(t_{2g}^5\ e_g^0)$ state for manganese, indicate the ability of the titled complex to be spin depolarizing, insulating, in its bulk (see also state (iii), figure 6b) to spin polarized currents.

Conclusion

We have shown for the first time using polarized neutron diffraction, density functional theory a visualization of quantum entanglement in view of the spin density distribution, electronic structure associated with the interactions of manganese in the titled complex. Our results for the titled complex is in support of the Einstein's theory of general relativity and show that the solution to the paradox of quantum formalism assumes leakage of hidden information pertaining to the entangled state during measurement. Finally, our results show the suitability of the complex for spintronic studies.

Acknowledgement

We are grateful for the financial support of the Swiss National Foundation for making this work possible. We would like to thank the Institut Laue Langevin (ILL) for access to the neutron diffraction facilities. We would also like to express thanks to the Condensed Matter Theory group and the Laboratory for Neutron Scattering and Imaging at the Paul Scherrer Institut.

Appendix A

For crystal data, experimental information for the synthesis of titled complex see reference [16].

Appendix B

For nuclear structure determination of the titled complex in the non-PND experiment, see reference [16]. The PND experiment[16] was performed with the same $2 \times 2.5 \times 5$ mm^3, 40 mg deuterated single crystal used in the non-PND experiment, on the D3 diffractometer at the Institut Laue Langevin, France. A measuring temperature of 1.6 K was used. A field of 9.6 Tesla applied perpendicular to the horizontal plane of the crystal, which had the [010] direction vertical. A Heusler crystal was used to monochromate the incoming neutron beam at a wavelength of 0.825Å with a polarisation of 0.975. A cryoflipper with 99% efficiency was used to change the direction of the polarization parallel and antiparallel relative to the direction of the applied field. As the complex crystallizes in the centrosymmetric space group R$\bar{3}$, the measured flipping ratio (R(**q**)) between the intensities $I^+(\mathbf{q})$ and $I^-(\mathbf{q})$ at the peak position of different Bragg reflections is defined by:

$$R(\mathbf{q}) = \frac{I^+(\mathbf{q})}{I^-(\mathbf{q})} = \frac{F_N(\mathbf{q})^2 + 2P^+F_N(\mathbf{q})F_M^{\perp}(\mathbf{q}) + F_M^{\perp}(\mathbf{q})^2}{F_N(\mathbf{q})^2 - 2P^-F_N(\mathbf{q})F_M^{\perp}(\mathbf{q}) + F_M^{\perp}(\mathbf{q})^2} \quad \text{(B.1)}$$

when the angle between the scattering vector and the magnetization direction is 90^0, the flipping ratio R(**q**) in (A.1) is re-defined as:

$$R(\mathbf{q}) = \frac{I^+(\mathbf{q})}{I^-(\mathbf{q})} = \frac{F_N(\mathbf{q})^2 + 2F_N(\mathbf{q})F_M(\mathbf{q}) + F_M(\mathbf{q})^2}{F_N(\mathbf{q})^2 - 2F_N(\mathbf{q})F_M(\mathbf{q}) + F_M(\mathbf{q})^2} \quad \text{(B.2)}$$

Journal of Physics: Conference Series **711** (2016) 012006 doi:10.1088/1742-6596/711/1/012006

Appendix C

Figure C1: Magnetization measurement on a powdered sample of the titled complex at 2 K , showing saturation at $5\mu_B$.

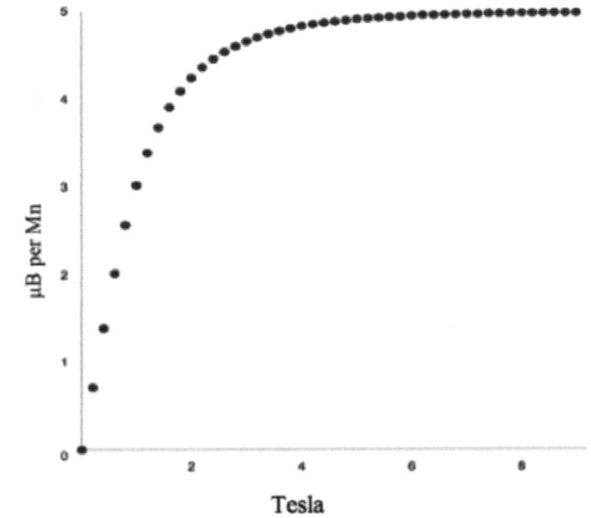

Figure C2: Temperature dependence of magnetic susceptibility performed on a powder sample of the titled complex at 500 Oe.

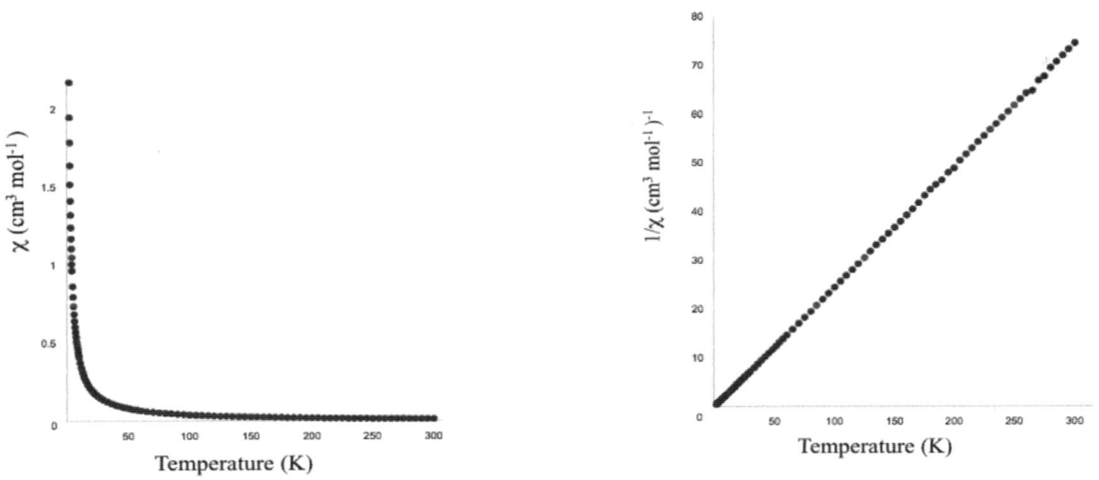

Journal of Physics: Conference Series **711** (2016) 012006 doi:10.1088/1742-6596/711/1/012006

Figure C3: High field EPR spectra (obtained at 280 K) for a powdered sample of the complex at 100.8 GHz, 201.6 GHz, 302.4 GHZ, 328.8 GHZ, 406.4 GHZ, showing no observed significant frequency dependence, an indication of entanglement for the electronic structure for manganese. The frequency independent splitting in the spectra cannot be described as an effect of electron g-factor anisotropy.

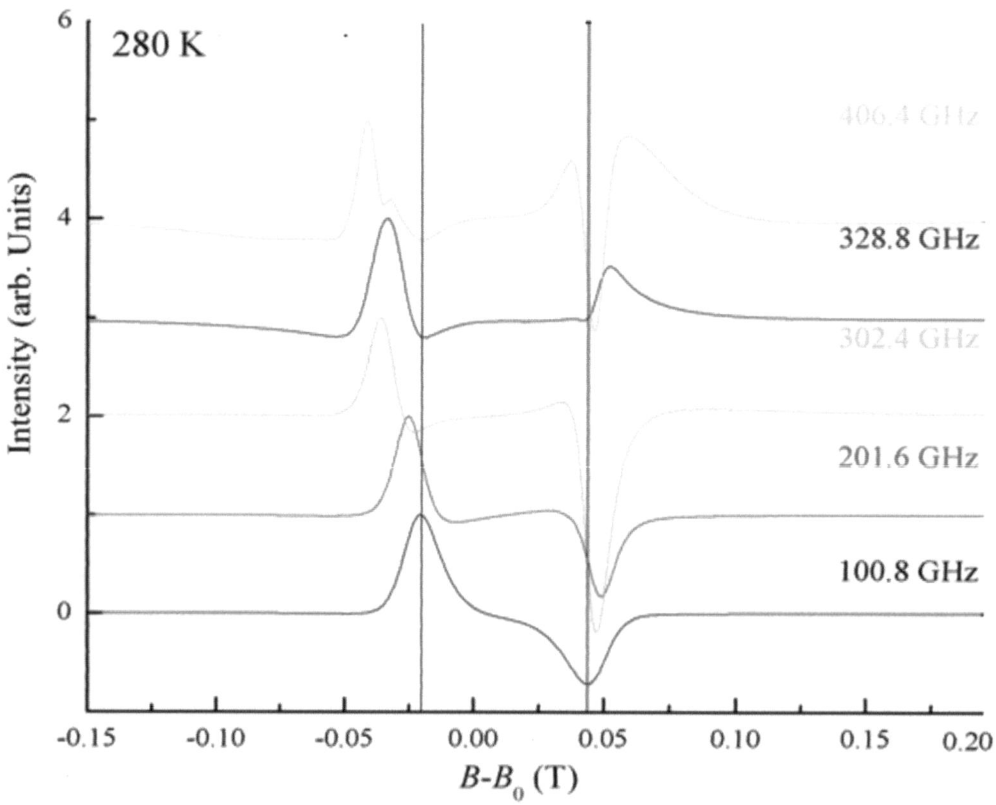

Reference

[1] Jaeger G. 2009 "Entanglement, Information, and the Interpretation of Quantum Mechanics": *The Frontiers Collection* XIV, p 307.

[2] Geralch W., Stern O. 1922 "Das magnetische Moment des Silberatoms" *Zeitschrift für Physik* 9, p. 353 – 355.

[3] Wilczek F. 1982 "Quantum Mechanics of Fractional-Spin Particles" *Phys. Rev. Lett.* 49(14), p. 957 – 959.

[4] Einstein A. 1905 "Ist die Trägheit eines Körpers von seinem Energieinhalt abhängig?" *Annalen der Physik* 13, 639 – 643.

[5] Minkowski H. 1908/9 "Raum und Zeit" *Physikalische Zeitschrift* 10, p. 75 – 88.

[6] Einstein A., Podolsky B., Rosen N. 1935 "Can Quantum-Mechanical Description of Physical Reality Be Considered Complete?" *Phys. Rev.* 47(10), p. 777 - 780.

[7] Einstein A. 1916 "The Foundation of the General Theory of Relativity" *Annalen der Physik* 354 (7), p. 769.

[8] Einstein A. 1915 "Die Feldgleichungen der Gravitation" *Sitzungsberichte der Preussischen Akademie der Wissenschaften zu Berlin* p. 844 – 847.

[9] Misner C. W., Thorne K. S., Wheeler J. A. 1973 "Gravitation" Chapter 34, p. 916, San Francisco, ed. W. H. Freeman.

[10] Schwarzschild K. 1916 "Über das Gravitationsfeld eines Massenpunktes nach der Einsteinschen Theorie" *Sitzungsberichte der Deutschen Akademie der Wissenschaften zu Berlin, Klasse für Mathematik, Physik, und Technik* p. 189.

[11] Schwarzschild K. 1916 "Über das Gravitationsfeld einer Kugel aus inkompressibler Flüssigkeit nach der Einsteinschen Theorie" *Sitzungsberichte der Deutschen Akademie der Wissenschaften zu Berlin, Klasse für Mathematik, Physik, und Technik* p. 424.

[12] Kerr R. P. 1963 "Gravitational field of a spinning mass as an example of algebraically special metrics" *Phys. Rev. Lett.* 11(5), p. 237.

[13] Penrose R. 1965 "Gravitational Collapse and Space-Time Singularities" *Phys. Rev. Lett.* 14(3), p. 57.

[14] Bardeen J. M., Carter B., Hawkins S. W. 1973 "The four laws of black hole mechanics" *Comm. Math. Phys.* 31(2), p. 161 - 170.

[15] Alperin H. A., Brown P. J., Nathans R., Pickart S. J. 1963 *J. Appl. Phys.* 34, p. 1182.

[16] Wallace W. A. 2013 *PhD Thesis* Laboratory for Neutron Scattering and Imaging, Paul Scherrer Institut, Switzerland.

[17] Heitler W., London F. 1927 *Z. Phy.* 44, 619.

[18] Pauling L. 1960 "The Nature of the Chemical Bond: An Introduction to Modern Structural Chemistry" Cornell University Press.

[19] Pauling L. 1931 *J. Am. Chem. Soc.* 53, 1367.

[20] Pauling L. 1931 *J. Am. Chem. Soc.* 53, 3225.

[21] Keffer F., Oguchi T., O'Sullivan W., Yamashita 1959 J. *Phys. Rev.* 115, 1553.

[22] Mason R., Varghese J. N. 1980 *Proc. Roy. Soc. Lond.(A)* 372, No. 1748, 1.

[23] Matthewman J. C., Thompson P., Brown P. J. 1982 *J. Appl. Cryst.* 15, 167.

[24] Bekenstein J. D. 1980 "Black-hole thermodynamics" *Physics Today* **1980**, 33, No. 1, 24 – 31.

[25] Delley B. *J. Chem. Phys.* **1990**, 92, 508.

[26] Delley B. *J. Chem. Phys.* **2000**, 113, No. 18, 7756.

[27] Einstein A. 1905 "Zur Elektrodynamik bewegter Körper" *Annalen der Physik* 17, 891, (English Translation: Jeffery, G. B., Perrett, W. 1923 "On the Electrodynamics of Moving Bodies").

Development of a ^{3}He nuclear spin flip system on an in-situ SEOP ^{3}He spin filter and demonstration for a neutron reflectometer and magnetic imaging technique

H Hayashida[1], T Oku[2], H Kira[1], K Sakai[2], K Hiroi[2], T Ino[3], T Shinohara[2], T Imagawa[4], M Ohkawara[5], K Ohoyama[5], K Kakurai[6], M Takeda[2,6], D Yamazaki[2], K Oikawa[2], M Harada[2], N Miyata[1], K Akutsu[1], M Mizusawa[1], J D Parker[1], Y Matsumoto[1], S Zhang[1], J Suzuki[1], K Soyama[2], K Aizawa[2], M Arai[2]

[1]Comprehensive Research Organization for Science and Society, Tokai, Ibaraki 319-1106, Japan
[2]J-PARC Center, Japan Atomic Energy Agency, Tokai, Ibaraki 319-1195, Japan
[3] High Energy Accelerator Research Organization (KEK), Tsukuba, Ibaraki 305-0801, Japan
[4] Central Research Laboratory, Hitachi, Ltd., 1-1, Omika-cho 7, Hitachi, Ibaraki 319-1292, Japan
[5] Institute for Materials Research, Tohoku University, Sendai, Miyagi 980-8577, Japan
[6] Quantum Beam Science Center, Japan Atomic Energy Agency, Tokai, Ibaraki 319-1195, Japan

E-mail: h_hayashida@cross.or.jp

Abstract. We have been developing a ^{3}He neutron spin filter (NSF) using the spin exchange optical pumping (SEOP) technique. The ^{3}He NSF provides a high-energy polarized neutron beam with large beam size. Moreover the ^{3}He NSF can work as a π-flipper for a polarized neutron beam by flipping the ^{3}He nuclear spin using a nuclear magnetic resonance (NMR) technique. For NMR with the in-situ SEOP technique, the polarization of the laser must be reversed simultaneously because a non-reversed laser reduces the polarization of the spin-flipped ^{3}He. To change the polarity of the laser, a half-wavelength plate was installed. The rotation angle of the half-wavelength plate was optimized, and a polarization of 97 % was obtained for the circularly polarized laser. The ^{3}He polarization reached 70 % and was stable over one week. A demonstration of the ^{3}He nuclear spin flip system was performed at the polarized neutron reflectometer SHARAKU (BL17) and NOBORU (BL10) at J-PARC. Off-specular measurement from a magnetic Fe/Cr thin film and magnetic imaging of a magnetic steel sheet were performed at BL17 and BL10, respectively.

1. Introduction

Polarized neutron scattering methods are effective for studying magnetic materials, and magnetic imaging techniques are also powerful tools to investigate and visualize magnetic structures in magnetic materials [1-5]. A ^{3}He neutron spin filter (NSF) is an effective device to provide a polarized neutron beam. A ^{3}He NSF has some advantages compared with a polarizing supermirror, another commonly used neutron polarizing device, because the ^{3}He NSF produces a polarized neutron beam with shorter wavelengths and cover a large solid angle [6]. The polarizing supermirror can also be made to cover a large solid angle by stacking multiple supermirrors. However the stacking precision of the supermirrors often creates a non-uniform distribution of neutron intensity, which is not an issue for the ^{3}He NSF. Moreover the ^{3}He NSF can be used as a π-flipper for a polarized neutron beam by simply flipping the ^{3}He nuclear spin using the nuclear magnetic resonance (NMR) technique. A ^{3}He NSF can thus provide a compact SEOP system which functions as both a neutron polarizing device and neutron spin flipper. Hence we attempted to combine ^{3}He nuclear spin flip system with the in-situ SEOP ^{3}He NSF that has been developed at J-PARC [7-12]. With the in-situ SEOP technique, the polarization of the laser must be reversed simultaneously with the ^{3}He spin, because a non-reversed laser destroy the polarization of the spin-flipped ^{3}He. The change of the polarity of the laser was accomplished using a half-wavelength plate installed after the circularly polarized laser. The details of the development of the ^{3}He nuclear spin flip system for the in-situ SEOP system is reported in section 2, and the demonstration of the system for off-specular reflection and magnetic imaging measurements are described in sections 3 and 4, respectively.

2. Nuclear spin flip system

Figure 1 (a) and (b) show the schematic diagram and photograph of the nuclear spin flip system using a half-wavelength plate. Circularly polarized laser light was emitted from the laser unit [12] and reflected by the quartz mirror, and a ^{3}He cell was placed after the quartz mirror. The laser unit consists of an air-cooled laser diode array (LDA) with output power of 30 W. Since the laser spectrum of the LDA is broad, a volume holographic grating (VHG) element is used to narrow the laser spectrum and match it to the Rb absorption line [12]. A half-wavelength plate (Thorlabs Inc., AHWP10M-980) was installed just downstream of the laser unit. The half-wavelength plate was set on an optical rail, and could be moved orthogonally to the laser beam direction, allowing us to change the polarity of the laser by inserting or removing the half-wavelength plate. A spectrometer (Thorlabs Inc., PAX5710IR1-T) was set after the ^{3}He cell position in order to measure the circularly polarized light incident on the ^{3}He cell. At first we measured the circular polarization without the half-wavelength plate, and we observed a right-handed polarized laser with a polarization of 97 %. Next, we inserted the half-wavelength plate, and after optimizing the rotation angle of the half-wavelength plate, we obtained a 97 % left-handed polarized laser light.

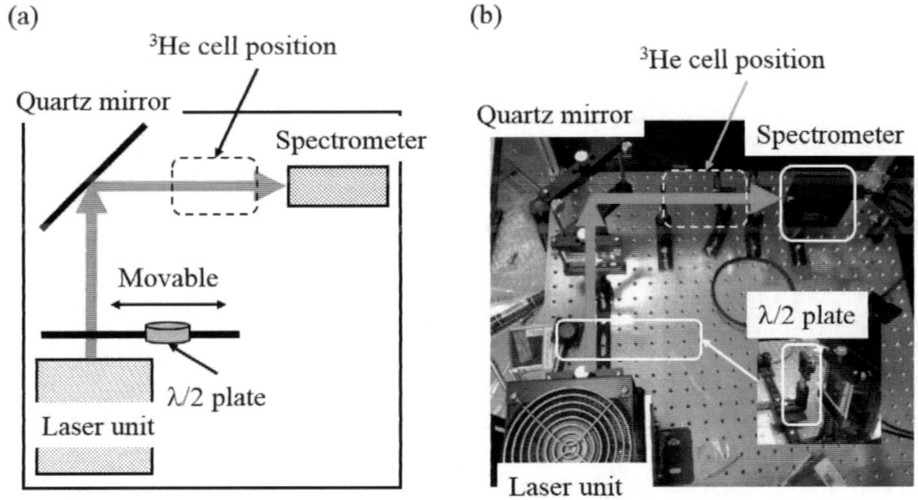

Figure 1. (a) the schematic diagram and (b) photograph of the ^{3}He nuclear spin flip system for the in-situ SEOP system.

After finishing the optimization of the half-wavelength plate, an in-situ SEOP pumping test was performed with the ^{3}He cell. An Adiabatic Fast Passage NMR (AFP-NMR) system was also installed to flip the ^{3}He polarization, and the polarization of ^{3}He nuclear spin was monitored via the voltage of the NMR signal V_{NMR}. The AFP-NMR system has two measurement modes, spin-flip mode and non-spin-flip mode, by changing the sweep function of current for the static field [11]. In this test, the ^{3}He polarization was monitored every 3 hours. Figure 2 shows the pumping result for the ^{3}He nuclear spin with the SEOP technique. A ^{3}He cell with a pressure length of 11.1 atm cm was used in this demonstration [11]. At about 40 hours after the start of laser pumping, the ^{3}He polarization was almost saturated, and the first ^{3}He nuclear spin flip on the in-situ SEOP was performed, shown as "Right → Left" in Fig. 2. The change to AFP-NMR spin-flip mode and the insertion of the half-wavelength plate were performed manually and not synchronized with each other, in the present state. V_{NMR} reversed with the ^{3}He nuclear spin as shown in Fig. 2, and V_{NMR} remained stable over 30 hours. After the first ^{3}He nuclear spin flip, the spin flip was performed an additional four times, as indicated by "Left → Right" and "Right → Left" in Fig. 2. In each spin flip test, V_{NMR} (^{3}He polarization) did not decay, and the ^{3}He nuclear spin flip system on the in-situ SEOP worked stably over 180 hours (7.5 days). These results show that the installed ^{3}He nuclear spin flip system provides good performance for the in-situ SEOP system. The maximum value of $|V_{NMR}|$ (absolute value of V_{NMR}) was about 34 mV, and it corresponds to a ^{3}He polarization P_{3He} of about 70 %. The P_{3He} was estimated using a calibration factor obtained by simultaneous measurement of neutron transmission and AFP-NMR in another experiment, which was previously reported in ref. [11].

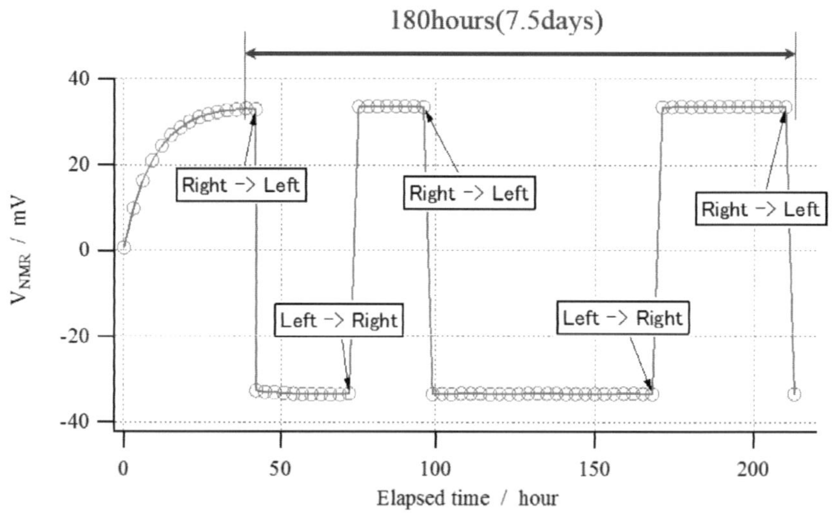

Figure 2. The results of the performance test of the ^{3}He nuclear spin flip system on the in-situ SEOP system.

3. Off-specular measurement

As a demonstration for neutron scattering, we installed the system in BL17 SHARAKU which is a neutron reflectometer at J-PARC. Normally, BL17 has a neutron spin polarizer and analyser consisting of polarizing supermirrors, and experiment modes using non-polarized or polarized neutrons can be selected [13]. Drabkin-flippers are installed before and after the sample position [14], and four patterns of neutron spin combination before and after the sample, (+, +), (-, -), (+, -) and (-, +), can be measured. Here, "+" and "-" are neutron spin states parallel or anti-parallel to a guide magnetic field, respectively. The analyser is fabricated from supermirrors (Fe/Si, $m = 4$) with a stack structure, covering a beam area of 50 mm in height and 100 mm in width with compact size (350 mm in length) and enabling off-specular and grazing incidence small-angle neutron scattering (GISANS) measurements. In such an analyser, however, it is impossible to stack the supermirrors at exactly equal angles, producing a small error that affects the off-specular and GISANS patterns. On the other hand, the ^{3}He NSF gives a smaller effect to the scattering patterns compared with stacked supermirrors. Hence BL17 requires the ^{3}He NSF as a neutron spin analyser.

The previous test of in-situ SEOP ^{3}He NSF has been performed and reported in ref. 10. However the ^{3}He nuclear spin flip system had not been installed in the in-situ SEOP system used in the past test. Therefore we could not select neutron spin states after the sample and could only measure two patterns of neutron spin states (-, -) and (+, +) as reported in ref. [10].

Figure 3 shows the photograph and schematic diagram of the current demonstration. In this demonstration, the pair of the Drabkin flipper and the analyser mirror that are normally placed between the 8th slit and 9th slit were exchanged for the in-situ SEOP ^{3}He NSF system with ^{3}He nuclear spin flip system. This system then enabled us to measure four patterns of neutron spin states (+, +), (+, -), (-, +) and (-, -) as shown in Fig. 3.

The ^{3}He cell with a pressure length of 11.1 atm cm with 35 mm in diameter and 55 mm in length was used. The cell was fabricated using GE180, and Rubidium is doped inside the cell. The ^{3}He nuclear polarization was monitored by AFP-NMR, and V_{NMR} was about 29 mV which corresponds to 60 % of polarization. That is lower than that of the off-line test (without neutron beam). The reason is thought to be due to the 25 Oe guide field generated from the guide coil placed about 300 mm upstream of the ^{3}He cell. However, the SEOP ^{3}He NSF system worked stably over the experiment time, and off-specular measurements could be performed.

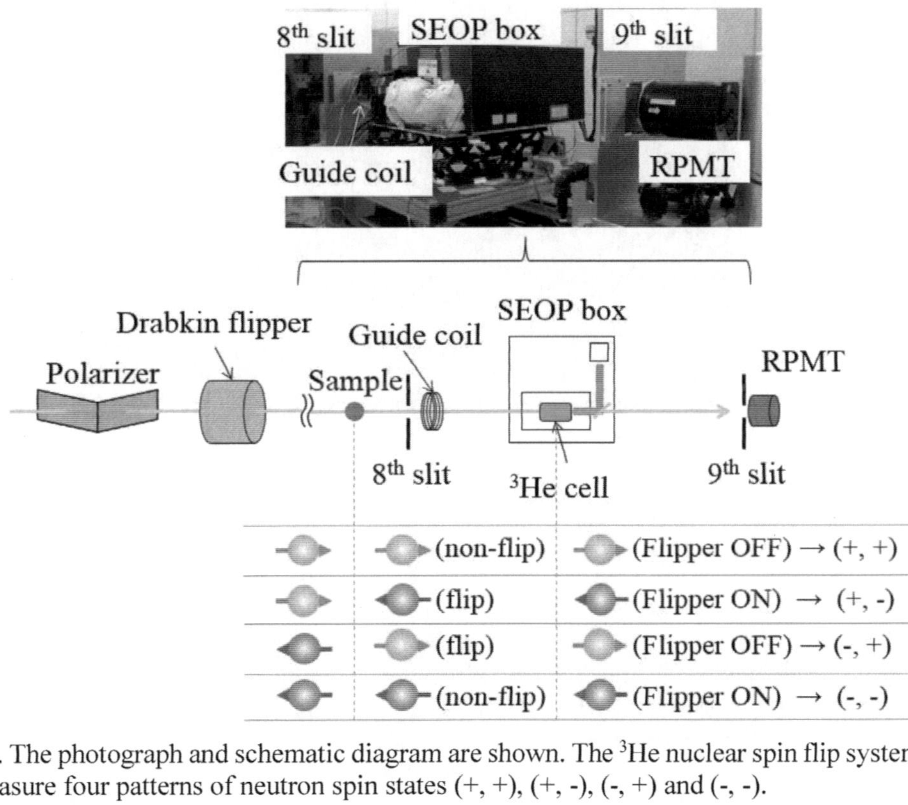

Figure 3. The photograph and schematic diagram are shown. The ^{3}He nuclear spin flip system enables us to measure four patterns of neutron spin states (+, +), (+, -), (-, +) and (-, -).

A Fe/Cr multilayer thin film with the giant magnetoresistance effect was used as a sample. In low magnetic field, the Fe layers create antiferromagnetic correlations that produce an off-specular reflection around $Q_z = 0.8$ nm^{-1}. This was the same sample used in the past experiment using in-situ SEOP at BL17 [10], and its features have already been reported in ref. [15]. In this experiment, the magnetic field applied to the sample was 200 Oe, and the off-specular reflections with four patterns (+, +), (+, -), (-, +)

and (-, -) were measured. Figure 4 shows the results of the off-specular measurements indicated on Q_x-Q_z map. Q_x and Q_z show the momentum transfers of the in-plane and depth structures on the sample, respectively. Off-specular reflections appear at $Q_x \neq 0$. In all off-specular patterns shown in Fig. 4, off-specular reflections are observed at about $Q_z = 0.8$ nm^{-1}. Moreover the off-specular patterns of (+, -) and (-, +) are almost identical. These features are consistent with the features reported in Ref. [15].

BL17 requires a large ^{3}He cell of about 150 mm in diameter, because the detector on BL17 has a 100 mm × 100 mm area. An improvement of the ^{3}He nuclear polarization is also necessary. We have recently started to develop a potassium-doped hybrid-cell to improve the ^{3}He nuclear polarization. Moreover we must adjust the pressure length of ^{3}He gas to match the peak intensity of the neutron beam at BL17. However, the ^{3}He nuclear spin flip system on the in-situ SEOP was demonstrated successfully and worked stably over 4 days of experiment time.

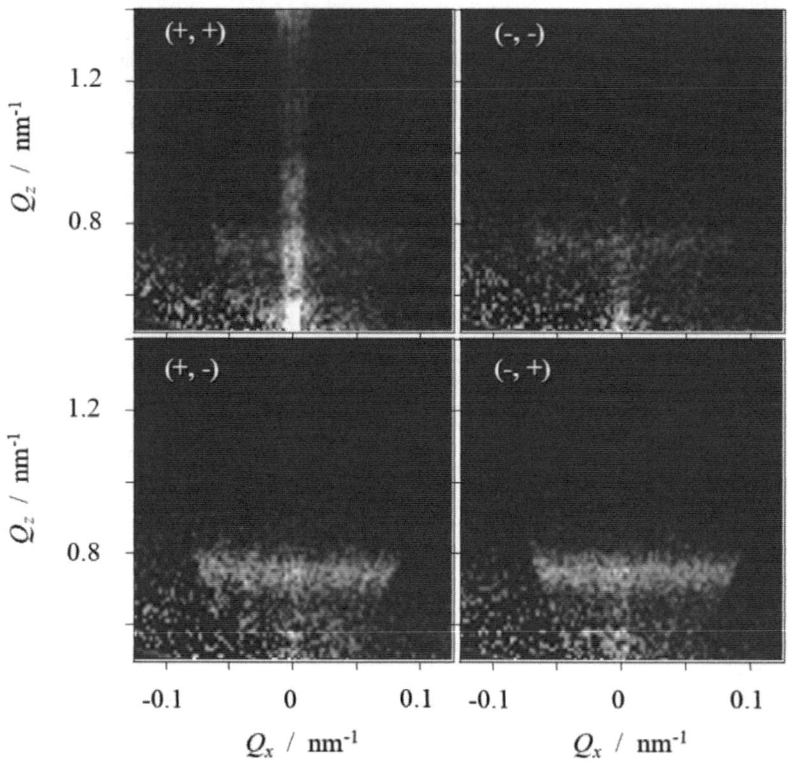

Figure 4. The results of the off-specular scattering measurement from Fe/Cr multilayer thin film with four neutron spin states.

4. Magnetic imaging

As another demonstration of the ^{3}He nuclear spin flip system using in-situ SEOP, we applied the system to a magnetic imaging technique at BL10 NOBORU (NeutrOn Beamline for Observation and Research Use) at J-PARC. When the neutron spin is depolarized entirely, it becomes impossible to obtain any information in a magnetic sample. For a magnetic sample with a strong magnetic field, the neutron spin becomes increasingly easy to depolarize. Hence using polarized neutrons with shorter wavelengths is very important, because shorter wavelength neutrons are not as easy to depolarize compared with longer wavelengths. The ^{3}He NSF is effective to provide polarized neutron with shorter wavelengths compared with a polarizing supermirror, therefore we started to apply the ^{3}He NSF to the magnetic imaging technique.

The photograph and schematic diagram of the experiment setup are shown in Fig. 5 (a). The in-situ SEOP ^{3}He NSF with the ^{3}He cell with pressure length of 17 atm cm was used as a neutron spin polarizer, and the ex-situ (not in-situ) one with that of 11.1 atm cm was used as a neutron spin analyser. The ^{3}He cell size was 35 mm in diameter and 55 mm in length for both. There was insufficient space for setting two in-situ SEOP systems in the BL10 experiment room, so in-situ SEOP was used only for the polarizer, including the ^{3}He nuclear spin flip system. An RPMT, a two-dimensional position sensitive detector consisting of a scintillator and position-sensitive photomultiplier tube, was placed after the analyser [16]. The sample was placed between the polarizer and the analyser.

A polarized neutron beam parallel/antiparallel to the guide field can be obtained without/with application of the ^{3}He nuclear spin flip on the in-situ SEOP polarizer. After rotating around a sample magnetic field $\boldsymbol{B}$, the neutron spins are analysed by the ex-situ SEOP. Then I_+ and I_- can be measured without and with application of the ^{3}He nuclear spin flip system, respectively. Neutron spin polarization P_{NS} can then be obtained from $P_{NS} = (I_+ - I_-) / (I_+ + I_-)$.

A magnetic steel sheet in the shape of a ring with a thickness of 0.35 mm was used as the sample as shown in Fig. 5 (b). About three-fourths of the sample was wrapped with a solenoid coil to apply and control a magnetic field to the sample. A neutron beam with a size of 15 mm × 15 mm was irradiated to the unwrapped portion of the magnetic steel sheet.

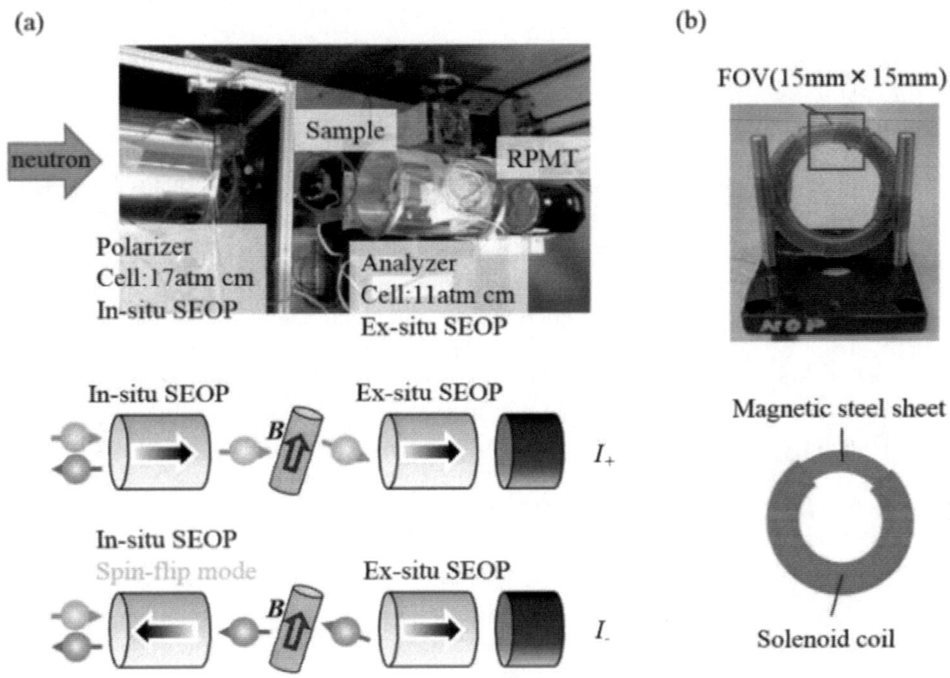

Figure 5. (a) Photograph and schematic diagram of the experiment setup, and (b) the ring shape of magnetic steel sheet.

The ^{3}He polarization of the polarizer was 40 %. Our laser unit has air cooling system [12], but in this experiment, the experiment room of BL10 was very hot and the air cooling system could not provide sufficient cooling. To prevent overheating, the laser power had to be reduced, leading to the lower ^{3}He polarization. However the ^{3}He nuclear polarization was kept stable over the experiment time. The ^{3}He NSF for the analyser was polarized before starting the magnetic imaging measurements. However the analyser was not in-situ SEOP, so the ^{3}He nuclear polarization decayed with time. The ^{3}He nuclear polarization before the start of the experiment was about 60 %, and during the 30 hours of the experiment, the polarization was reduced to about 40 %, where these percentages were estimated from the AFP-NMR measurements. The polarization of the direct neutron beam (without sample) was also reduced. Figure 6 shows the change of the neutron beam polarization, where the circles and triangles show the polarization before and after the magnetic imaging measurements.

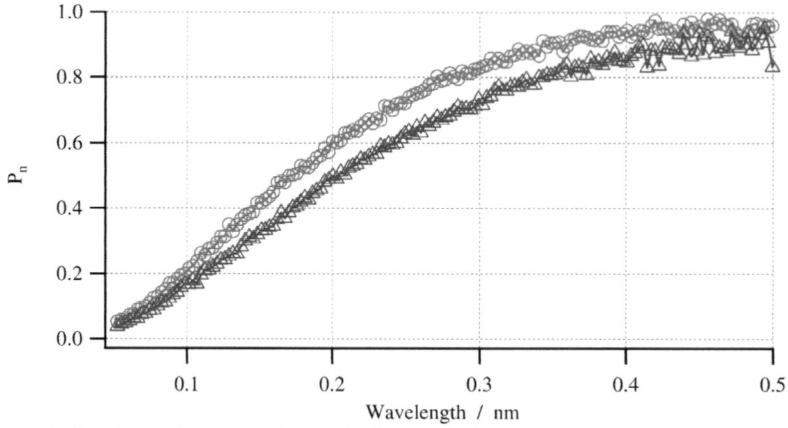

Figure 6. The polarization of neutron beam before (circle) and after (triangle) the magnetic imaging measurements.

Figure 7 shows the results of the magnetic imaging measurement of the magnetic steel sheet. Figure 7 (a) and (c) are two-dimensional (2D) images of neutron polarization at a neutron wavelength of 0.27 nm (1 pixel = 0.5 mm). Initially, we applied a current of $I = 1.42$ A to the solenoid coil on the sample, and the 2D image of the neutron polarization is shown in Fig. 7 (a). Next, the current was reduced to 0 A, and the resulting 2D image is shown in Fig. 7 (c). In both of these images, the neutron polarization P_{NS} is normalized by the polarization of the direct beam for each measurement time $P_{ND}(t)$, where t is the measurement time. Then we obtain the normalized neutron polarization from $P_N = P_{NS}/P_{ND}(t)$. Figure 7 (b) shows the wavelength dependence of P_N extracted from the square region indicated in Fig. 7 (a). Similarly, figure 7 (d) shows P_N from the square region indicated in Fig. 7 (c). The square regions in Fig. 7 (a) and (c) are at the same position on the sample. The wavelength dependence of P_N is caused by a rotation of neutron spin around a magnetic field. The oscillation cycle shows an integral of magnetic field of a neutron path, and depolarization of neutron spin leads to a decay of the amplitude of P_N [3, 11]. Both the 2D images and the wavelength-dependent polarization show drastically different patterns for the $I = 1.42$ A and $I = 0$ A cases. In this measurement, only the magnetic component perpendicular to the quantization axis of the neutron spin was measured, therefore it is impossible to determine the intensity and direction of magnetic field quantitatively. However, these results clearly show that the distribution of magnetization in the magnetic steel sheet changed with different applied magnetic field.

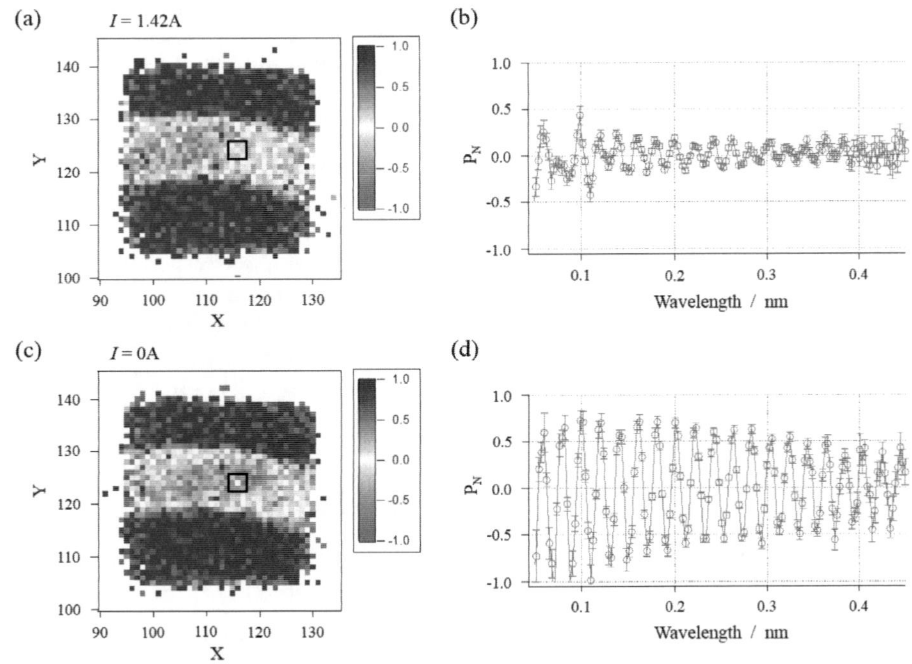

Figure 7. The results of the magnetic imaging measurements. (a) and (c) are 2D images of P_N from I = 1.42 A and I = 0 A at λ = 0.27 nm, respectively. (b) and (d) show the wavelength dependence of P_N of same regions shown in (a) and (c). The pixel size of the 2D maps (a) and (c) is 0.5 mm.

In this demonstration, the ^{3}He nuclear spin flip system on the in-situ SEOP worked stably during the entire experiment time of 30 hours, and magnetic imaging using polarized neutrons at shorter wavelengths was successfully performed. However neutron polarization of the direct beam was not high, for example P_N = 0.2 at λ = 0.1 nm. We are planning to use the polarized neutron with λ = 0.05 nm for a strong magnetic field sample, for example a stacked magnetic steel sheet and electrical motor. Therefore big improvements are necessary for all components of the in-situ SEOP ^{3}He NSF system as follows: 1) a high ^{3}He nuclear polarization of more than 80 %, 2) a high pressure length of ^{3}He cell (of more than 45 atm cm), and 3) a laser unit with power greater than 100 W in a compact size of less than 150 mm in length, 150 mm in width and 100 mm in height. Moreover, as excessive distance between the sample and detector leads to blurring of the 2D image due to beam divergence, an extremely compact in-situ SEOP system (with a length less than 300 mm) for the neutron spin analyser is needed. Therefore the extremely compact analyser system is very important for the magnetic imaging technique. Accomplishment of these requirements is not easy, but we must continue to develop each component.

5. Conclusion

A ^{3}He nuclear spin flip system for an in-situ SEOP ^{3}He NSF has been developed using a half-wave plate. Pumping and nuclear spin flip tests were performed, and the system worked stably over 180 hours (7.5 days) with several ^{3}He nuclear spin flips.

A demonstration of the ^{3}He nuclear spin flip system for the in-situ SEOP was performed at the neutron reflectometer SHARAKU (BL17) at J-PARC. The in-situ SEOP ^{3}He NSF was used as a neutron spin analyser. A ^{3}He cell with a pressure length of 11.1 atm cm, a diameter of 35 mm, and a length of 55 mm was employed, and a Fe/Cr multi-layered thin film was used as a sample. Off-specular measurements were performed, and the ^{3}He nuclear spin flip system enabled us to measure four neutron spin states, (+, +), (+, -), (-, +) and (-, -). The applied magnetic field was 200 Oe, and off-specular patterns were observed at Q_z = 0.8 nm^{-1}. The off-specular patterns with neutron spin flipped (+, -) and

(-, +) were almost the same. These results are consistent with features reported in ref. [15]. The demonstration was successfully performed, however we need to continue some development to optimize the system for use at BL17, including improvement of the ^{3}He nuclear polarization to 80 %, enlargement of the cell size to cover 100 mm × 100 mm areas, and optimization of the ^{3}He gas pressure length to the neutron wavelength distribution at BL17.

Next, we applied the in-situ SEOP to a magnetic imaging technique. The in-situ SEOP using a ^{3}He cell with pressure length of 17 atm cm was used as a neutron spin polarizer, and the ex-situ SEOP with pressure length of 11.1 atm cm was used as a neutron spin analyser. By using the ^{3}He nuclear spin flip system, the intensities of both parallel and anti-parallel spin polarized neutrons could be measured, and we could obtain the neutron spin polarization P_N. A magnetic steel sheet with solenoid coil was used as the sample. By changing the current in the solenoid coil, magnetic field applied to the sample could be controlled, and magnetic imaging measurements with two current conditions, $I = 1.42$ A and $I = 0$ A, were performed. 2D images of P_N were obtained, and different P_N patterns between the two conditions were observed. Moreover the observed wavelength dependences of P_N for a specific area of the sample were also obviously different for each condition, showing again that the magnetization patterns between the two conditions were altered. This demonstration was also successfully performed, however many improvements are still needed including a higher ^{3}He nuclear polarization of more than 80 %, a high pressure length of more than 45 atm cm, and an extremely compact in-situ SEOP system with length less than 300 mm for the neutron spin analyser.

Acknowledgement
This work was supported by the Quantum Beam Fundamentals Development Program, Japan Ministry of Education, Culture, Sports, Science, and Technology (MEXT). This work was performed under the Japan Atomic Energy Agency (JAEA) program of project use (Proposal No.: 2012P0803).

References
[1] Kardjilov N, et al., 2008 *Nat. Phys.* ***4*** *399.*
[2] Piegsa F M, et al., 2008 *Nucl. Instr. and Meth. A* ***586*** *15.*
[3] Shinohara T, et al., 2011 *Nucl. Instr. and Meth. A* ***651*** *121.*
[4] Hayashida H, et al., 2011 *Nucl. Instr. and Merh. A* ***634*** *S90.*
[5] Hiroi K, et al., *JPS Conference Proceedings, in press.*
[6] Walker T G, et al., 1997 *Rev. Mod. Phys.* ***69*** *629.*
[7] Ino T, et al., 2009 *Physica B* ***404*** *2667.*
[8] Kira H, et al., 2011 *J. Phys.: Conf. Ser.* ***294*** *012014.*
[9] Arimoto Y, et al., 2011 *Physica B* ***406*** *2439.*
[10] Hayashida H, et al., 2014 *J. Phys.: Conf. Ser.* ***528*** *012020.*
[11] Sakai K, et al., *J. Phys.: Conf. Ser.* ***528*** *(2014) 012016.*
[12] Oku T, et al., *JPS Conference Proceedings, in press.*
[13] Takeda M, et al, 2012 *Chinese J. Phys.* ***50*** *161.*
[14] Hayashida H, et al., 2013 *Physics Procedia* ***42*** *130-135.*
[15] Takeda M, et al., 1995 *Physica B* ***213&214*** *248.*
[16] Hirota K, et al., 2005 *Phys. Chem. Chem. Phys.* *7 1836.*

SEOP polarized ^{3}He Neutron Spin Filters for the JCNS user program

Earl Babcock[1*], Zahir Salhi[1], Tobias Theisselmann[1], Denis Starostin[1], Johann Schmeissner[1,2], Artem Feoktystov[1], Stefan Mattauch[1], Patrick Pistel[3], Aurel Radulescu[1], Alexander Ioffe[1]

[1] Jülich Centre for Neutron Science (JCNS) at the Heinz Maier-Leibnitz Zentrum (MLZ), Lichtenbergstaße. 1, 85747 Garching Germany

[2] *current* National Research Nuclear University MEPhi, Kashira Hwy, 31, Moscow, Russia, 115409

[3] Forschungszentrum Jülich GmbH ZEA-1, Leo-Brandt-Straße, 52425 Jülich Germany

E-mail: * `e.babcock@fz-juelich.de`

Abstract. Over the past several years the JCNS has been developing in-house applications for neutron polarization analysis (PA). These methods include PA for separation of incoherent from coherent scattering in soft matter studies (SANS), and online polarization for analysis for neutron reflectometry, SANS, GISANS and eventually spectroscopy. This paper will present an overview of the user activities at the JCNS at the MLZ and gives an overview of the polarization ^{3}He methods and devices used. Additionally we will summarise current projects which will further support the user activities using polarised ^{3}He spin filters.

1. Introduction

SEOP has long been proposed as a method for in-situ polarization of ^{3}He neutron spin filters (^{3}He NSF) and has now been realised at many centres worldwide [1–5]. Offline, or laboratory polarised ^{3}He NSF have also been used and refined to great success [6, 7]. However while being a simpler, albeit very robust, installation on the neutron instrument, these systems require continual maintenance to refresh the cell's polarization and often lead to the necessity of a ^{3}He service group which may or may not be available for all centres. Because of this, the long term strategy of the JCNS has been to focus on in-situ SEOP polarizers which are customised and integrated to each particular application, leading to long term stability, independence and scalability [8].

We currently have one long-time running in-situ polarizer developed for the MARIA magnetism reflectometer [9]. This system has undergone many upgrades to the lasers, magnetic cavity and heating system which will be the focus of an upcoming paper. It is used as the prototyping bed for other in-situ polarised applications for our institute such as for analysis on the SANS diffractometers KWS-1 and KWS-2 and for for incident beam polarization on a thermal time-of-flight spectrometer TOPAS [10–12]. The methods for standard polarized reflectometry and SANS using ^{3}He NSF analysis are "routine" as such instrument capabilities have been explored [13, 14]. The goal of much of our program is to further develop the methods, and to increase the available Q-coverage of the analyzer systems over what is currently available to add to the scientific applications of the methods which were summarized in [15]. Finally the

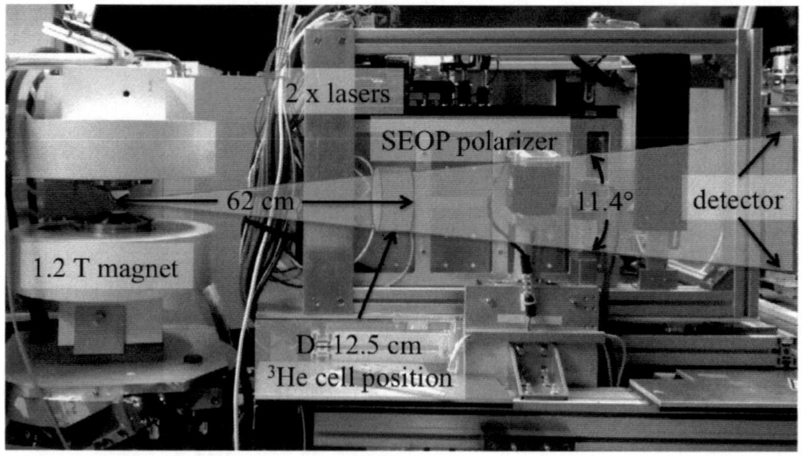

Figure 1. A picture of the current revision of the MARIA analyzer with the neutron/laser shielding box removed. The field geometry of the box is orthogonal to the 1.2 T sample magnet field (2.2 T with reduced pole-gap) in order to prevent magnetic coupling of the two systems. This along with the angled front yokes allow a closer approach of the front of the ^{3}He spin filter cell to the sample area, about 55 cm, thus increasing the angular-coverage or Q-range to over 11° or 0.23Å^{-1} for 4.5Å neutrons and a D=12.5 cm, L=8 cm cell. Above the box the two VBG narrowed lasers are mounted which are directed onto the cell with 3 mirrors.

system for TOPAS will also build upon the experience learned from the other devices, largely the MARIA analyzer, but will polarize the incident beam, which for the case of TOPAS's energy range and focusing guide geometry would not be possible with a super-mirror type device [12]. This paper will describe the complete MARIA analyzer and it's applications, as well as the KWS-1 and KWS-2 systems that are in production, giving concrete examples from user experiments. Then we will discuss the prototyping under way to provide in-situ polarization for each application to support the long term user needs.

2. MARIA polarization ^{3}He analyzer

Our current online SEOP system for MARIA, which will serve as the model for many other JCNS instruments such as KWS1, KWS2 and TOPAS, has many innovations. These innovations have been arrived upon over a long time with experience and testing of online SEOP polarizers for neutron instrumentation beginning with work on PF1B at the ILL and at ISIS [4, 16]. The beginning system for MARIA was essentially a further optimised version of these predecessors. However the system has continually evolved into its current state. The refinements cover every major subsystem of the device including: Python-based NMR and control program with web-GUI interface, upgraded laser systems, all-electric heating systems, new optimised magnetic field geometry and neutron spin transport. A detailed description of all of these devices will be the subject of a future instrument publication, here we will summarise each of these components and the overall performance characteristics of the final device. A picture of the MARIA analyzer system is given in figure 1.

NMR free induction decay (FID) and adiabatic fast passage (AFP) flipping are both available on all JCNS ^{3}He NSF systems. The FID allows one to monitor the ^{3}He polarzation over time and characterize the optical pumping parameters such as the spin-up time constant and cell lifetime. Our FID system conserves the phase information of the ^{3}He precession and thus also allows us to probe whether the ^{3}He is in the up or down state. It is a single-coil pulse and receive system similar to that of [17] however using a house developed analogue switch and signal conditioning amplifier similar to [18]. The AFP allows the ^{3}He NSF to double as a high performance, wide area neutron flipper as well [19]. The AFP program is an all digitally generated pulse similar to that in [20], and amplified using a house built power amplifier with a 120 V_{pp} output from 5 kHz up to 100 kHz. For the case of in-situ optical pumping via SEOP, the handedness of the circular polarization, and therefore the alkali-metal polarization, must be in phase with the direction of the ^{3}He spin. For this purpose we use liquid crystal variable wave retarders, which can be switched between 1/2 λ and 1 λ retardance by applying a TTL pulse to a house built analogue liquid crystal (LC) wave plate driver. For our system there are 3 mirrors at approximately 45° to steer the laser beams into the cell, each of which can cause a slight rotation of the plane of polarization. To compensate for this, we found it most robust to use the LC as a switchable 1/2 to 1 λ retarder and a fixed 0-order quartz 1/4 λ at 795 nm retarder set to give net 1/4 λ and 3/4 λ retardances in conjunction with the effect of the mirrors to provide the resulting right and left handed polarizations of the pumping lasers. All the NMR devices and the switching of the LC retarders are accomplished with a program written in Python which has a web-browser interface that can be accessed remotely over a web server or via IP protocols. The program controls a 1.25 MHz Lab-view DAQ card using the National Instruments DAQmx 8.0 driver library which runs in either Linux or Windows.

Our laser systems have also been upgraded. Formerly diode array bars (DAB) with external cavities formed by plane-reflection diffraction gratings in a Littrow-type configuration were used [21] as narrow-band optical pumping light has been shown to provide the best SEOP performance. However continuation of this work has found a new method using chipped volume Bragg gratings, CVBG, to provide higher absolute ^{3}He polarizations [22]. Additionally CVBG narrowed lasers give higher laser output for the same power laser diode, and contain much fewer optical elements making the cavity more compact, stable and robust for long term operation when used in in-situ ^{3}He polarizer applications. A photo of the new laser cavity is shown in figure 2. In this photo the laser is shown mounted inside a newly designed compact box for mounting the laser and cavity optics, plus external connections to the current cables and cooling water. This box also provides environmental protection and increased laser security. Two of these lasers are used due to the large 0.85 litter volume of the ^{3}He NSF cell used on MARIA, however other applications using smaller volume cells will use only one.

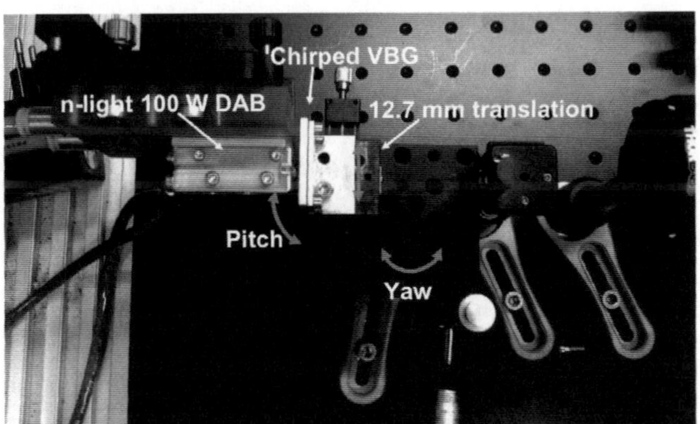

Figure 2. A picture of the current volume Bragg grating narrowed laser configuration on MARIA. This optical setup has now been adapted to fit in a custom made enclose box to enclose and mount all the optics and provide connections to power cable and cooling water.

As can be seen in figure 1 the magic box-style magneto-static cavity has also been updated and improved. The principle of operation remains the same to that of prior magic-boxes for online SEOP polarization [4,9] however the front set of magnetic "core-yokes" have been rotated by 45°. This modification accomplishes two things: one, the angled plates help shift the homogeneous region of the box forward towards the sample area, and two, they reduce the area of the opening

Journal of Physics: Conference Series **711** (2016) 012008 doi:10.1088/1742-6596/711/1/012008

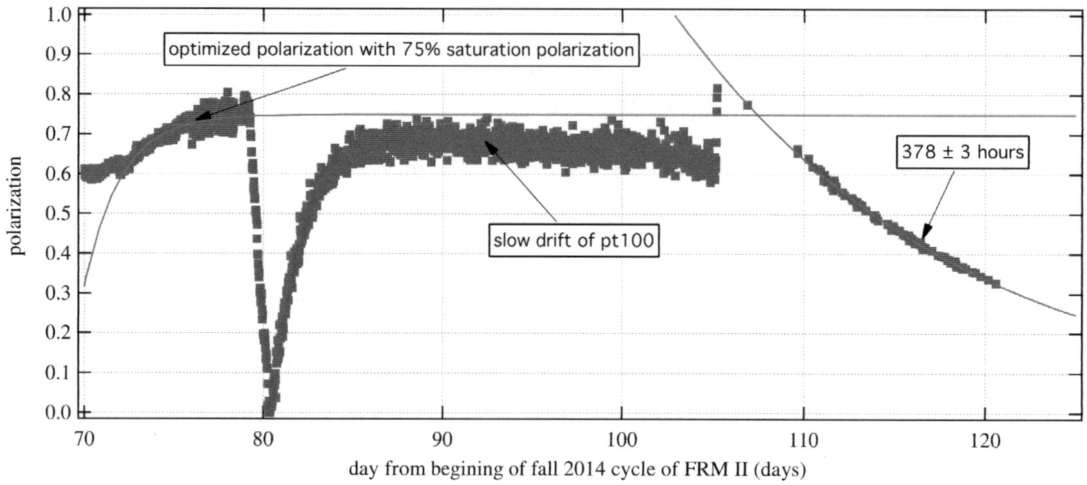

Figure 3. A graph of the MARIA analyzer ^{3}He polarization during the 18 October to 18 December 2014 cycle of the FRMII, days 75 to 105 on the graph. At the beginning of the shown data the system was re-optimised. At day 80 the system polarization was intentionally inverted with respect to the optical polarization and was polarised through 0. The slow degradation is caused by a long term degradation of the pt100 temperature sensor leads/connection. At the end of the data a room temperature T_1 lifetime measurement of the cell taken during a reactor pause can be seen. The measured T_1 is the same as was measured in laboratory conditions for this cell.

in the cavity on the side towards the sample magnet providing for better magnetic shielding from the field of it. This allows the cell to be moved closer to the sample, providing larger angular coverage for the same size cell. The cell used in MARIA, called Idefix [23], is 12.5 cm diameter and 8 cm long. In the revised magic box the front face of Idefix is about 27 cm from the magix box entrance, and 55 cm from the MARIA sample position providing a $\pm 5.7°$ angular coverage around the direct beam. Lastly, in comparison to the first revision of the MARIA analyzer [9] here we placed the magic box B_0 field orthogonal to that of the sample magnet. In the prior parallel-field geometry, the magnetic circuit of the magic-box was found to couple to the anti-parallel return flux of the sample magnet, causing zeo-field regions near the plane of the scattered neutron beam, leading to neutron beam depolarization before the ^{3}He NSF cell. This happens because the μ-metal of the magic box, when placed sufficiently close to the sample magnet, provides a high permeability return path for the sample magnet's flux that competes with the magnet's yoke. When explored with a hall probe, and also confirmed via finite element modelling of the coupled system, this essentially causes mirror images of the sample magnet coils to be formed inside the magic box above and below the symmetry axis of the beam, at sufficiently high sample magnetic fields, this then resulted in low field areas that partially depolarize the scattered neutron beam passing through these regions.

The orthogonal field geometry solves the problem of the coupling of the two magnetic systems. This insures that neutron spin transport is maintained at any sample filed, and the magnitude of the field at the position of the ^{3}He cell is stable to $1\text{x}10^{-4}$ or better (NMR frequency shifts of < 5 Hz) for any sample field from 1 mT up to the maximum of 1.2 T (with the standard pole-shoes installed). In this geometry the magic box can be placed very close to the sample magnet, however due to the requirement to rotate and tilt the sample environment, and mechanics for a movable beam stop, plus radiation shielding/laser light shielding there is a space of approximately 10 cm between the perimeter of the sample magnet and the front face of the magic box. Therefore a short permanent magnet guide field, parallel to the sample magnet field is attached to the entrance window of the radiation-laser shield in order to maintain a minimum vertical neutron guide field of at least 10 G for all sample fields. The vertical field of this permanent magnet guide field overlaps sufficiently with the orthogonal field of the magic box such that 10 G is maintained everywhere and the neutron polarization rotates adiabatically for all neutron wavelengths used on MARIA which are ≥ 4.5 Å .

An un-anticipated benefit of this new construction is not only that the cell becomes closer to the sample position increasing its Q-range, but that the magnetic homogeneity has also improved. In the current cavity a recent room temperature T_1 measurement of the Idefix cell used for MARIA resulted in a value of 378±3 hours, the same value as was obtained in the laboratory for this cell [23]. This is an improvement compared to the prior simple rectangular magic box (L=80 cm W=30 cm H=40 cm with the field along the 40 cm direction and oriented parallel to the sample field) where the same cell obtained a 300 − 320 hour lifetime depending on the sample field used. Since the currently measured lifetime in the MARIA system is within errors equal to the T_1 lifetime measurement of the same cell in laboratory conditions, the improvement is entirely attributed to improvement in the magnetic field homogeneity and improved shielding of the revised "angled-plate" magic-box magnetic cavity.

The overall performance is good and highly suitable for difficult user experiments. The SEOP ^{3}He NSF cell uses a hybrid Rb-K mixture to achieve high optical pumping efficiency [24, 25] and has been prepared in house [23]. The Idefix cell has an internal diameter of 12.5 cm, length of 8 cm and a ^{3}He lifetime of 380 hours in laboratory conditions using a 6-coil, 2 m diameter and 2.5 m long, high homogeneity holding field. The details and preparation of the Idefix cell, as well as those of several other cells produced by us can be found in [23]. As stated earlier, inside the new MARIA magic-box this cell has a room temperature lifetime of 378 ±3 hours; this is more than sufficient to achieve high polarizations using the SEOP method.

The heating system has been upgraded from a heat-gun based, forced air approach to now use static air heated by electric cartidge heaters mounted in aluminium heat sinks. These stainless steel electric cartridge heaters (220 V CSS-series from Omega Engineering) now used for heating have not been found to decrease expected ^{3}He polarization, or room temperature T_1 lifetime. There is indeed some magnetic components to these heaters, namely the nickel plated lead wires, however in our installation, the two heaters are placed on ether side of the oven, inside the before mentioned aluminium heat-sinks, about 10 cm away from the cell, with the wires exiting away from the cell position. The aluminium heat sinks share are grounded to limit pickup of electrical noise by the NMR FID system.

Running at an oven temperature of approximately 185°C a saturation polarization of 70% is achieved in this cell. For this temperature in a hybrid cell [24] the spin exchange rate is slow; 50 hours for the Idefix cell used. The resulting polarization is highly stable at around 70% but this saturation polarization could be increased to 75% or more by by going to oven temperatures on the order of 20°C to 30°C hotter which would produce faster spin-up time constants. However, we will first install active temperature stabilization for the external enclosure of the device before pushing the parameters to the eventual limits.

In figure 3 the long term operation of the system can bee seen. In this particular data, it was run for over 160 days continuously, achieving a maximum 75% ^{3}He polarization value after a re-optimisation of cell temperature and laser tuning at around day 70 which was directly before the FRM II winter 2014 cycle that began at day 75 on the graph. During this 160 day span several user experiments have been performed. The time span of figure 3 particularly shows the ^{3}He polarization over the winter 2014 30-day half cycle from November 18th to December 18th 2014 and several weeks of a T_1 measurement over the reactor break from Dec 18th 2014 to Jan 13th

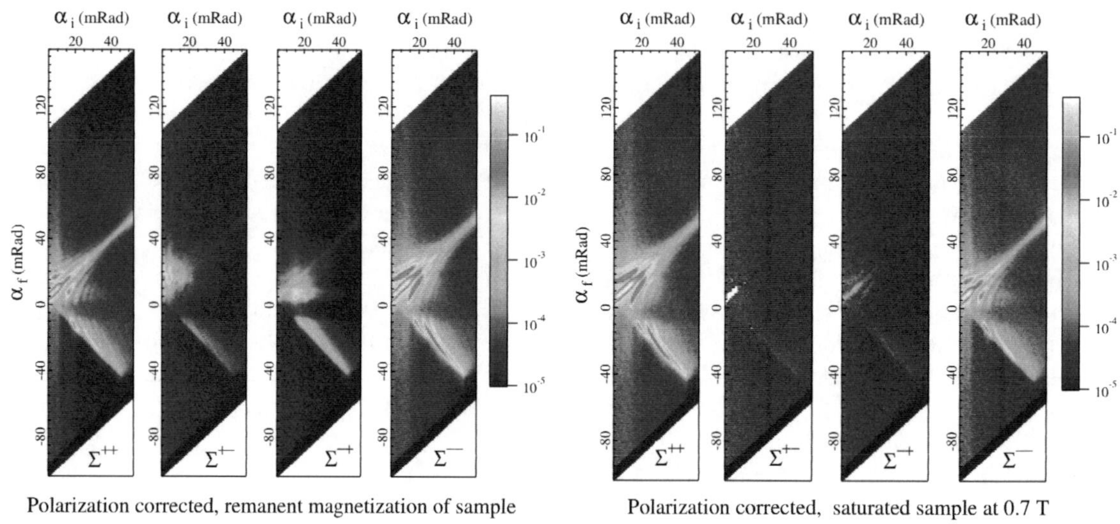

Figure 4. Polarised off-specular reflectivity data from the $^{nat}Fe^{58}Fe$ superlatice grating sample. An asymmetry in the spin-flip cross sections, graphs Σ^{+-} and Σ^{-+}, can be seen in the measurement at remanent magnetisation that disappears in the data to the right where an external 0.7 T field is applied. Note, the remnant field of the sample was anti-parallel to the applied field from the sample magnet. Close inspection of the diagonal spectral scattering shows a dynamic range of over 1x10^{-4} is achieved in the polarised mode. The data displayed are the normalised scattering cross-sections which have been corrected for the PA efficiencies.

2015. At the beginning of the data, the cell polarization was inverted and re-polarised through 0 (day 80) followed by a maximum of about 70% which slowly decayed due to a long term downward "drift" in temperature due degradation of the connections to our pt100 temperature sensor. Following this data the lasers and heating were shut off and a measurement of the cell T_1 lifetime can be seen. Subsequent to this data, the temperature sensor has been replaced by one with longer all platinum leads such that the connection of the pt100 to the temperature controller cable is outside the oven solving the problem of long term drift. For the spring 2015 FRM II reactor cycle the ^{3}He polarization was maintained constant for over 30 days during user experiments without any tuning of the lasers or oven temperature.

3. Examples from Reflectometry, GISANS and SANS data

In this section we will highlight some user measurements made with the ^{3}He NSF program at JCNS. As a test of polarization analysis on MARIA, a $^{nat}Fe^{58}Fe$ superlatice grating was measured. This sample had 10 repetitions of a ^{nat}Fe layer followed by a ^{58}Fe layer each of 0.1 μm thickness with a 10 μm period grating lithographically etched on the top surface. The sample was produced by the K. Temst group at KU Leuven [26], and was on loan from A. Wildes at the Insitut Laue-Langevin (ILL). Figure 4 shows α_f vs. α_i (final angle vs. incident angle) plots for the four polarization states at 1 mT (remanent) and 0.7 T (saturated) sample field. As can be seen in figure 4, rich off-specular scattering data can be obtained over a very large angular range and the specular reflectivity measured shows over a 1x10^4 dynamic range for the full polarization analysis mode. Before the 1 mT data the sample was magnetised anti-parallel to the neutron guide field and thus the remanent magnetization was anti-parallel to the neutron polarization for this data. One can see many orders of diffraction from the grating in the off-

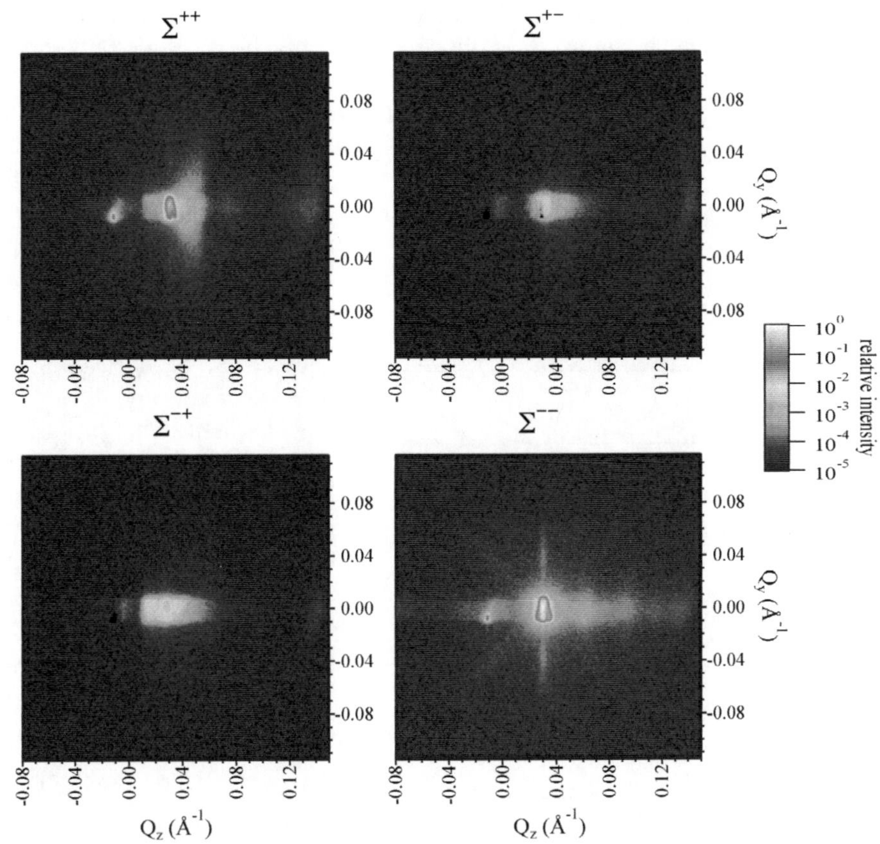

Figure 5. Full polarization analysis GISANS detector images from the magnetic multilayer (E. Kentzinger, et. al. MLZ proposal no. 10010) are shown. From the non spin-flip scattering in the Σ^{++} and Σ^{--} images, interfacial roughness is observed and from the spin-flip scattering in the Σ^{+-} and Σ^{-+} images, the GISANS from magnetic fluctuations is seen. The data was taken with 6 Å neutrons, the Q-values have a range of 0.230Å^{-1} for both the out of plane Q_z and in plane Q_x scattering corresponding to an angular coverage of $\pm 5.7°$. The detector was at an angle of 2° and the sample at an angle of 1° with respect to the direct beam. The data shown are normalized PA efficiency corrected cross sections.

specular data, further a spin-flip asymmetry can be observed in the remenant-magnetization measurement which disappears under magnetic re-orientation by the applied 0.7 T field.

The data shown in figure 4 have been corrected for the cell transmission and the polarizer, polarizer-flipper, and analyzer efficiencies following the methods which were clearly presented in [13]. In our case however, the procedure is somewhat simplified because the AFP ^{3}He flipper is an ideal neutron flipper [19], thus removing the inefficiency correction of the second flipper from the relations given in [13]. Further, for the current case of the in-situ polarizer on MARIA, the steady-state polarization means that time-dependent analyzing-power or transmission corrections are not needed when the system is in nominal operation. Work to model this data with the BornAgain project (i.e. software for simulations of GISANS and off-specular data using the DBW approximation) [27] as a model system is underway.

Another difficult application, polarization analysis with GISANS, is also possible. User experiments of magnetic multilayers (E. Kentzinger, et. al.) at the JCNS for the MLZ proposal no 10010 are shown in figure 5. Spin-flip scattering in the GISANS data show the magnetic fluctuations in the sample and in the non spin-flip scattering one observes the interfacial roughness, GISANS is a unique way of separating these two effects. This work is a continuation of studies on such magnetic multilayers [28, 29], however the polarization analysis GISANS experiment is performed now for the first time. Again, the data shown in figure 5 has been normalized for the polarized ^{3}He NSF cell transmission and corrected for the polarizer efficiency, polarizer-flipper efficiency and ^{3}He NSF polarization efficiency again as above using the basic method of [13].

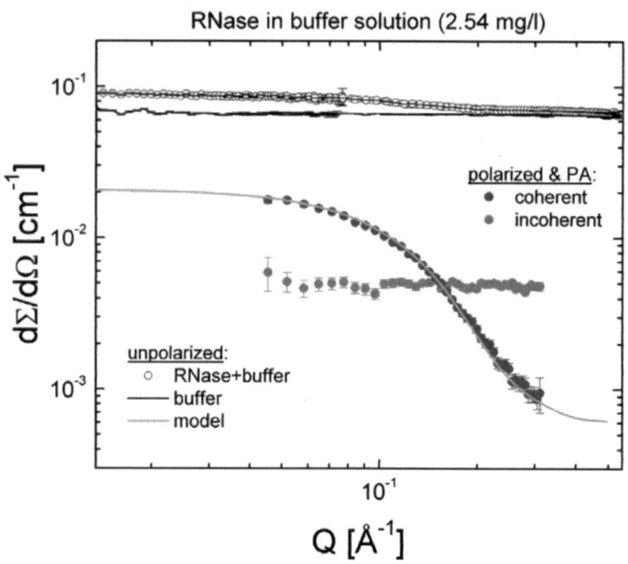

Figure 6. Polarised SANS data from an experiment by J. Fischer and R. Biehl. The unpolarised data for the Ribonuclease sample plus buffer solution and the buffer solution alone are at the top in black. One can see the scattering from the sample is only 10-20% above that of the buffer solution. Using polarization analysis the coherent scattering from the Ribonuclease sample is clearly obtained to a level 100 times below the total scattering, and 10 times below the level of innocent scattering of the Ribonuclease sample itself. The model-fit is also shown.

Lastly, a long proposed use of ^{3}He NSF is for the separation of coherent scattering from incoherent scattering scattering in soft matter studies [11, 15]. Work is continuing towards this goal, and more experiments have been done to further develop and refine the methods. Here we show one particular example of a user experiment. In this case a Ribonuclease sample is measured in a buffer solution, the data is shown in figure 6 [30]. One can see in the black curves at the top of the figure 6, corresponding to the unpolarised measurements of the sample in buffer and the buffer alone, that there is a large contribution from the buffer solution and the sample's own incoherent background. With the application of polarization analysis, seen in the colored data points, the coherent signal from the Ribonuclease can be clearly extracted to a level 100 times lower than the total scattering of the sample plus buffer, and to a factor of 10 times below the incoherent scattering of the Ribonuclease sample itself. In figure 6 the coherent form factor is compared to a calculated form factor based on the Ribonuclease A crystal structure including a hydration shell with 4% increased solvent density matches the absolute intensity of the coherent scattering [31]. This experiment was conducted with the provisional off-line polarised ^{3}He NSF cells and system described in [11] which is limited to a maximum Q-range of 0.3 $Å^{-1}$ for a 4.5 Å incident neutron beam.

4. Planned in-situ devices for KWS1, KWS2 and TOPAS

For the two small angle scattering instruments at the JCNS in the MLZ, two very different in-situ polarizers are planned. The two instruments have different scientific focuses, KWS2 [15] being mainly soft matter for the separation of incoherent background, whereas KWS1 focuses on magnetism research [10]. Soft matter studies will put highest priority on maximum Q-range whereas magnetism requires operation of the system in proximity to a high field sample magnet as on MARIA.

Therefore the system being built for KWS1 looks very similar to that on MARIA, but at 50% scale due to space constraints in the sample area for the sample magnet plus ^{3}He NSF analyzer system together. Further the mechanical design has been updated to be simpler and the length further shortened to 32 cm which was calculated to be the minimum possible for this style of magnetic cavity and still to have good performance. A diagram of the KWS1 design as optimised using a 3D FEM calculation which includes a model of the sample magnet is given in figure 7. The total length of the installation including the sample magnet will be under 1 m along the neutron beam to fit in the available sample environment space on KWS1. The magnetic cavity is currently under construction after which it will be tested first in the laboratory, and then on KWS1 using offline polarised ^{3}He NSF cells.

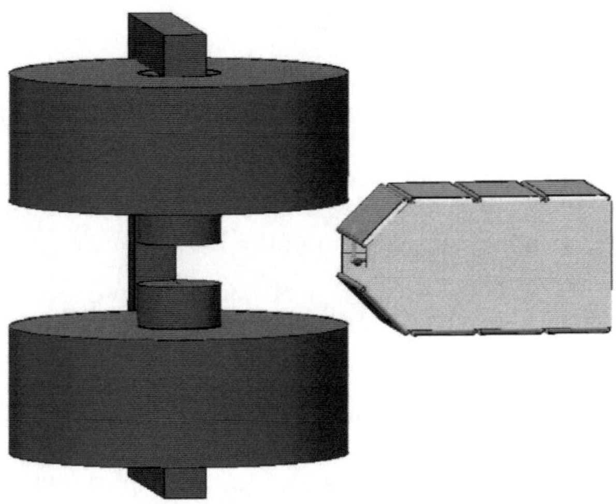

Figure 7. A diagram of the KWS1 magnetic system, similar to that on MARIA, as used to calculate the magnetic fields and compatibility. This system is designed to provide a similar angular coverage, up to $\pm 6.5°$ or $Q \leq 0.16Å^{-1}$ using 4.5 Å incident neutron wavelength. The total length will be under 1 m along the neutron beam. Not shown is the laser and optics which will be mounted above the cavity as on MARIA.

The KWS2 analyzer will look much different as the goal is to place the sample as close as possible to the cell to increase the possible Q-range. Additionally the system must minimise the total length so as not to increase the effective minimum sample to detector distance because again, as for KWS1, the system will be placed in the sample environment area, between the sample and the detector tank. The KWS2 system will use a very compact solenoid cavity which has been described in previous proceedings [11]. The total length of this solenoid is 18 cm and allows a sample to be placed within 3 cm of the front of the ^{3}He NSF cell. This magnetic cavity has been constructed and tested for ^{3}He polarization lifetime: the lifetime of a test cell was measured to be about 400 hours, giving a magnetic relaxation term of approximately 1000 hours for the cavity. This value is again more than sufficient for in-situ polarization. The oven and heating system have also been prototyped. This system will undergo further testing, optimisation and mechanical refinement before eventual user operations.

The TOPAS in-situ polarizer will be the neutron polarizer for the incident beam. Thus a much more conservative approach is taken. A simple rectangular SEOP magic-box style magnetic cavity will be used very similar to those in [4, 9]. The box design has been resized for the smaller beam diameter required for an incident beam of about D=6 cm. The resulting cavity is 30 cm tall by 20 cm wide and 50 cm long. the field will be along the 30cm direction and be co-linear with the guide fields of TOPAS to prevent depolarization of the very fast 0.7 Å minimum neutron wavelength to be used there. Two VBG narrowed DAB lasers as mentioned

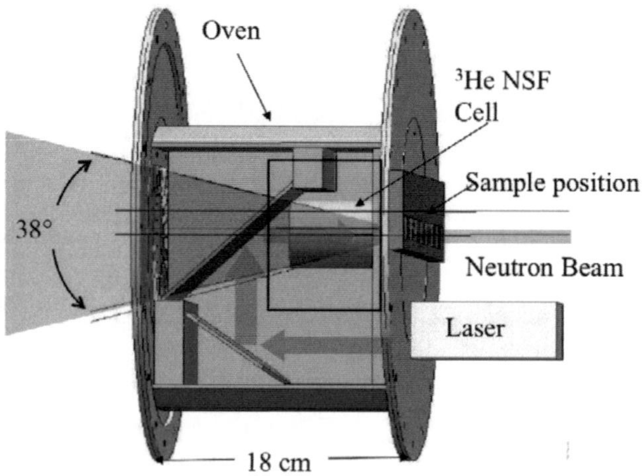

Figure 8. A diagram of the KWS2 magnetic system, SEOP oven and optics. The main components of the system have been prototyped and tested. Further mechanical optimisation and neutron testing will be conducted before routine use.

before will be used, one for the front 7.5 cm and the other for the back 7.5 cm of the 15 cm long cell needed to polarised the TOPAS incident beam. The magnetic cavity is nearing completion and will be tested soon. Several ^{3}He NSF cells with 3 bar ^{3}He and 6 cm diameter and 15 cm length have been produced and tested for lifetime, lifetimes in excess of 200 hours are achieved.

For the TOPAS analyzer, an offline polarised banana-shaped cell will be used in conjunction with a magic-PASTIS magnetic system. This device is described more fully in a separate paper in this same proceedings issue [32].

5. Conclusions and acknowledgements

The program for in-situ SEOP ^{3}He NSF at the JCNS at the MLZ has made many advancements. Several systems are in construction or testing phases and a working prototype has been in use on MARIA for several reactor cycles and corresponding user experiments. In total, four unique in-situ SEOP polarizers are currently planned for the JCNS instrument suite. These include the readily available MARIA analyzer, analyzer for KWS1 (in production) and polarizer for the incident beam on TOPAS (in production) all of which utilize a magic-box rectangular mu-metal magnetic capacitor geometry, and KWS2 (in testing) which uses an ultra compact solenoid geometry.

We thank colleagues at other neutron centres who have been involved in fruitful exchanges of ideas: T.R. Gentile and W.C. Chen at NIST, and S. Boag at ISIS. Further we thank K. Temst of KU Leuven for use of the ^{nat}Fe^{58}Fe superlatice grating sample, J. Fischer and R. Biehl for permission to show the Ribonuclease polarised SANS data curves, and E. Kentzinger for permission to show the polarised GISANS data images from the sample of magnetic multilayers.

References

[1] Hayashida H *et al* 2014 *J. Phys.: Conf. Ser.* **528** 012020 (proc. of NOP&D 2013)
[2] Jones G *et al* 2006 *Physica B-Condensed Matter* **385-86** 1131–1133
[3] Tong X *et al* 2012 *Rev. Sci. Inst.* **83**
[4] Boag S *et al* 2009 *Physica B-Condensed Matter* **404** 2659–2662
[5] Chupp T *et al* 2007 *Nuclear Instruments and Methods in Physics Research Section A* **574** 500–509
[6] Chen W *et al* 2014 *J. Phys. Conf. Ser.* **528** 012014
[7] Andersen K *et al* 2009 *Physica B-Condensed Matter* **404** 2652–2654
[8] Babcock E and Ioffe A 2011 *Physica B-Condensed Matter* **406** 2448–2452
[9] Babcock E, Mattauch S and Ioffe A 2011 *Nuclear Instruments and Methods in Physics Research Section A* **625** 43–46
[10] Feoktystov A *et al* 2015 *J. Appl. Cryst.* **48** 61–70 URL `http://dx.doi.org/10.1107/S1600576714025977`

[11] Babcock E *et al* 2013 *Physics Procedia* **42** 154–162
[12] Voigt J, Babcock E and Brueckel T 2010 *J. Phys. Conf. Ser.* **211** 012032
[13] Wildes A 1999 *Rev. Sci. Inst.* **70** 4241–4245
[14] Krycka K, Chen W, Borchers J, Maranville B and Watson S 2012 *J. Appl. Cryst.* **45** 546–553
[15] Ioffe A *et al* 2012 *Chinese J. Phys.* **50** 137–154
[16] Babcock E *et al* 2009 *Phys. Rev. A* **80** 033414
[17] Parnell S, Woolley E, Boag S and Frost C 2008 *Meas. Sci. & Tech.* **19** 045601
[18] Babcock E 2005 *Spin-exchange optical pumping with alkali-metal vapors* Ph.D. thesis University of Wisconsin-Madison section 3.6.1
[19] Babcock E *et al* 2007 *Physica B-Condensed Matter* **397** 172–175
[20] McKetterick T *et al* 2011 *Physica B-Condensed Matter* **406** 2436–2438
[21] Babcock E, Chann B, Nelson I and Walker T 2005 *Applied Optics* **44** 3098–3104
[22] Chen W, Gentile T, Ye Q, Walker T and Babcock E 2014 *J. Appl. Phys.* **116** 014903
[23] Salhi Z, Babcock E, Pistel P and Ioffe A 2014 *J. Phys. Conf. Ser.* **528** 012015
[24] Chen W, Gentile T, Walker T and Babcock E 2007 *Phys. Rev. A* **75** 013416
[25] Babcock E, Chann B, Walker T, Chen W and Gentile T 2006 *Phys. Rev. Lett.* **96** 083003
[26] Temst K 2015 KU Leuven personal page for K. Temst http://www.kuleuven.be/wieiswie/en/person/00016278
[27] Durniak C, Pospelov G, Van Herck W and Wuttke J 2015 BornAgain *Software for simulating and fitting X-ray and neutron small-angle scattering at grazing incidence* http://www.bornagainproject.org
[28] Kentzinger E *et al* 2007 *Physica B-Condensed Matter* **397** 43–46
[29] Kentzinger E, Ruecker U, Toperverg B, Ott F and Brueckel T 2008 *Phys. Rev B* **77**
[30] Fischer J, Biehl R, Hoffmann B and Richter D 2013 *European Biophysics J. with Biophysics Lett.* **42** S65
[31] Svergun D *et al* 1998 *Proc. of the National Academy of Sciences of the United States of America* **95** 2267–2272
[32] Salhi Z *et al* 2015 *J. Phys. Conf. Ser.* **this issue**

PNCMI 2014 IOP Publishing
Journal of Physics: Conference Series **711** (2016) 012009 doi:10.1088/1742-6596/711/1/012009

VERITAS: a high-flux neutron reflectometer with vertical sample geometry for a long pulse spallation source

S. Mattauch[1], A. Ioffe[1], D. Lott[2], A. Menelle[3], F. Ott[3], Z. Medic[4]

[1] JCNS at MLZ, Forschungszentrum-Jülich GmbH, 85747 Garching, Germany

[2]Helmholtz Zentrum Geesthacht, 21502 Geesthacht, Germany

[3]Laboratoire Léon Brillouin CEA/CNRS, CEA Saclay, 91191 Gif sur Yvette, France

[4] Faculty of Sciences, University of Novi Sad, Trg Dositeja Obradovića 3, 21000 Novi Sad, Republic of Serbia

E-mail: s.mattauch@fz-juelich@de

Abstract.

An instrument concept of a reflectometer with a vertical sample geometry fitted to the long pulse structure of a spallation source, called "VERITAS" at the ESS, is presented. It focuses on designing a reflectometer with high intensity at the lowest possible background following the users´ demand to investigate thin layers or interfacial areas in the sub-nanometer length scale. The high intensity approach of the vertical reflectometer fits very well to the long pulse structure of the ESS. Its main goal is to deliver as much usable intensity as possible at the sample position and be able to access a reflectivity range of 8 orders of magnitude and more. The concept assures that the reflectivity measurements can be performed in its best way to maximize the flux delivered to the sample. The reflectometer is optimized for studies of (magnetic) layers having thicknesses down to 5Å and a surface area of $1 \times 1 cm^2$. With reflectivity measurements the depth-resolved, laterally averaged chemical and magnetic profile can be investigated. By using polarised neutrons, additional vector information on the in-plane magnetic correlations (off-specular scattering at the μm length scale, GISANS at the nm length scale) can be studied. The full polarisation analysis could be used for soft matter samples to correct for incoherent scattering which is presently limiting neutron reflectivity studies to a reflectivity range on the order of 10^{-6}.

1. Introduction

The European Spallation Source (ESS) will provide neutron pulses with a width of τ=2.84ms and a repetition rate of 14Hz. Though the average flux at the ESS (with its TDR moderator) will be practically equal to the one at the ILL, the time structure of the neutron beam allows for a drastic gain in intensity for time-of-flight instruments due to the 25 times higher peak intensity of the ESS [1]. The natural resolution of the instrument using the full pulse width is τ/L and defined by the choice of the instrument length L. On the other hand, the artificial narrowing of the pulse width by pulse shaping choppers allows to increase the resolution at the cost of intensity. This opens an exciting opportunity to design a next-generation reflectometer to meet the increasing demand and anticipated scientific challenges. The research topics that will benefit from this reflectometer comprises a wide range of scientific disciplines, ranging from thin film magnetism and novel topological phases in confined geometries, the functionality and properties of hybrid materials in the field of soft and hard matter to the structural biology of membrane proteins. Though the proposed vertical sample geometry excludes the examination of liquid-liquid or liquid-gas interfaces, it nevertheless provides sufficient advantages for soft matter samples that do not require such interfaces and can be measured on the vertical reflectometer with the appropriate sample environment without compromises. Full polarisation analysis will allow measurements of those sample types that were not possible before by making it possible to correct for incoherent background.

The instrument concept presented here, called "VERITAS" the ESS, focuses on designing a reflectometer with high intensity and low background following the high demand of the users to investigate thin layers or interfaces in the sub-nanometer length scale. The high intensity approach of the vertical reflectometer fits very well to the long pulse structure of the ESS. Its main goal is to deliver as much usable intensity as possible to the sample position and be able to access a reflectivity range of 8 orders of magnitude and more.

The concept assures that the reflectivity measurements can be performed in its best way to maximize the flux delivered to the sample. The reflectometer is optimized for studies of (magnetic) layers having thicknesses down to 5Å and a surface area of 1x1cm^2. With reflectivity measurements the depth-resolved, laterally averaged chemical and magnetic profile can be investigated. By using polarised neutrons, additional vector information on the in-plane magnetic correlations (off-specular scattering at the μm length scale, GISANS at the nm length scale) can be studied.

The instrument will furthermore be capable to work with a higher wavelength resolution down to 1%. Depending on the operational mode of the instrument, different detector configurations will be used. The detector area will be highly configurable and optimised for the different needs of the specular, off-specular and GISANS-modes.

The design of the vertical reflectometer is based on well-tested components that will be very robust and bear no unpredictable risks for a reliable operation.

1. General philosophy: relaxed Q-resolution machine

The proposed instrument is primarily designed for the investigations of thin interfaces from several nm down to the sub nm range. The main goal is therefore to deliver as much usable intensity as possible at the sample position to be able to access a reflectivity range of 8 orders of magnitude and more. Fig. 1 shows the specular reflectivity curves simulated for a thin Fe-layer for a perfect and relaxed Q-resolutions, respectively. Comparing the two cases it can be noticed that only the minima of the interference pattern are slightly smeared out in the latter case, thus the resolution for the measurement can be drastically relaxed for thin interfacial structures without any loss of information. With a relaxed resolution the neutron intensity on the sample is increased to obtain a detectable signal from an extremely small amount of the scattering material of a thin layer, particularly if it is required to measure the reflectivity up to high Q values.

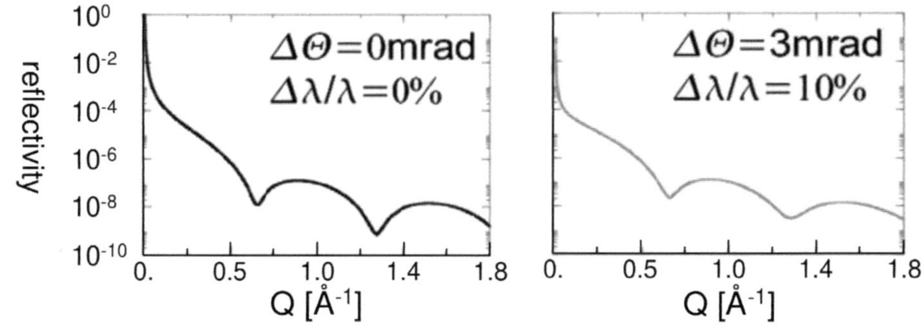

Fig. 1: Simulated reflectivity curve of 10Å thin Fe layer on a Ag substrate for ideal resolution/collimation and for relaxed wavelength resolution/collimation.

The pulse width τ and the instrument length *L* from the moderator to the detector impose physical limits on the main instrument parameters – the natural λ-resolution and the wavelength band Δλ, are both directly determined by the choice of τ and L:

$$\frac{\Delta\lambda}{\lambda} \propto \frac{\tau}{T} \propto \frac{\tau}{L} \qquad \Delta\lambda \propto \frac{1}{L} \tag{1}$$

(T- the time of flight of neutrons from the moderator to the detector). To achieve a resolution of 10% for τ=2.86ms at the ESS and to use a maximum of neutrons from the spectrum centered around a wavelength of 3Å, the instrument length L is fixed at about 36m. When L is chosen no further relaxation of the wavelength resolution is possible while the increase in the resolution can be achieved by a shortening of the pulse length. In turn the repetition rate of 14Hz and the choice of instrument length defines the wavelength band of the instrument to Δλ=8Å.

2. General instrument layout

The general layout of the instrument is depicted in Fig. 2, the schematic diagram of the instrument is shown in Fig. 3. The overall length up to the detector position is 36m allowing a wavelength resolution of 10% (see above) by making use of the full length of the ESS pulse.

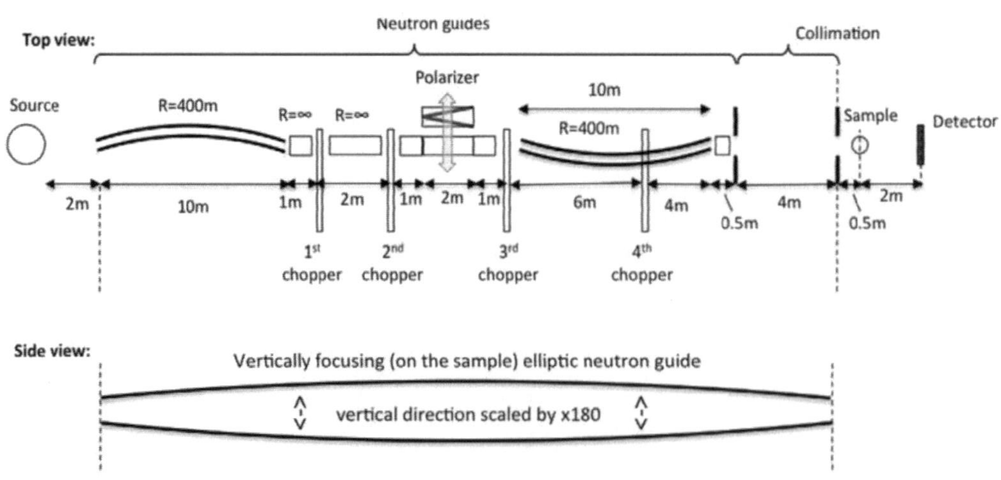

Fig. 2: The top and side view of VERITAS. The moderator is on the left hand side.

2.1. Neutron guide design:
The instrument guide is S-shaped and made of two curved neutron guides (R=400m) with a 7m long straight neutron guide at the inflection point (see Fig. 2). The use of the S-shaped neutron guide prevents the direct line-of-sight of the primary and secondary radiation sources. The movable parts (choppers, polariser changer, etc.) are positioned downstream of the neutron beam beyond the biological shielding and thus can be freely accessed during the operation of the ESS. The basic guide parameters are listed in Table 1. The neutron guide width of 30mm is chosen to allow the complete filling of the phase space of the guide for a moderator width of 10cm. To increase the incident neutron intensity for reflectivity measurements on small samples, the incident neutron beam is focused along the vertical direction onto the sample using a focusing elliptical neutron guide. A comparison of the intensities at the sample position (for a sample size of 1x1 cm^2) of such a setup to a straight guide with a cross section of 3 x 12 cm^2 and mirrors with m=2 coating shows a clear increase in the beam intensity by at least a factor of 4 for the elliptical focusing option (see Fig. 3).

An elliptical guide (see Table 1 for the shape) with the same length and m=3 coating for the last 4m will be used.

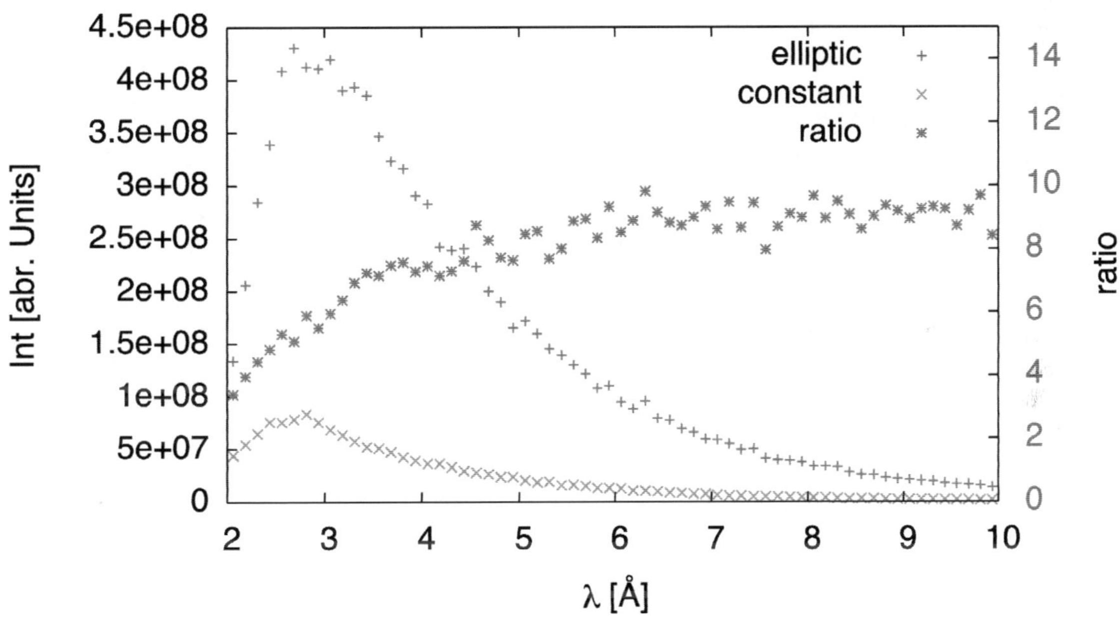

Fig. 3: Comparison of the intensity at the sample position with a sample size of 1x1cm² for elliptical and constant cross section neutron guides.

2.2. Chopper design:
The time-distance diagram for the low resolution mode is shown in Fig. 4a. The first chopper at 13m is the band selection chopper selecting an 8Å broad wavelength band. The band from 2 to 10Å will provide the highest intensity. The wavelength band can be selected arbitrarily in a range between 2 and 32Å. This is important in the case one wants to adjust the reciprocal space to a specific scientific question, e.g. separating the off-specular scattering signal from the direct beam or in the GISANS mode for separating reflections from each other. The additional choppers ((2) at 15m, (3) at 19m and

(4) at 25m) serve as frame overlap choppers to prevent contaminations of very high wavelength neutrons of more than 50Å.

	gth				Exit height	Exit width	ng		Coating left	Coating right
1	10.0m	+400m	95.8mm	30mm	194.5mm	30mm	m=3.0	m=3.0	m=3	m=4
2	1.0m	∞	194.5mm	30mm	198.5mm	30mm	m=3.0	m=3.0	m=3	m=3
3	2.0m	∞	198.5mm	30mm	202.7mm	30mm	m=3.0	m=3.0	m=3	m=3
4	4.0m	∞	202.7mm	30mm	202.5mm	30mm	m=3.0	m=3.0	m=3	m=3
5	6.0m	-400m	202.5mm	30mm	180.5mm	30mm	m=3.0	m=3.0	m=4	m=3
6	4.0m	-400m	180.5mm	30mm	145.5mm	30mm	m=3.0	m=3.0	m=4	m=3
7	0.5m	∞	145.5mm	30mm	140.0mm	30mm	m=3.0	m=3.0	m=3	m=3
8	4.0m	∞	140.0mm	30mm	54.0mm	30mm	m=3.0	m=3.0	absorber	absorber

Table 1: Basic guide design parameters. To avoid the depolarization of the neutron beam, all guides after the polarizing cavity (from section 4) are coated with non-magnetic supermirrors. Geometric parameters of the elliptic guide were selected to maximize the brilliance transfer, while keeping the guide's cross-section reasonably small for a practical chopper design.

In order to achieve a high wavelength resolution for the proposed design, the natural pulse width has to be reduced according to Eq. (1). Technically the sub-pulse duration will be defined by a pulse-shaping chopper installed at 13m from the source. Each sub-pulse will provide neutrons arriving at the detector during a certain time interval (time sub-frame) and covering a certain wavelength band (wavelength sub-frame), see Fig. 4b. For the optimal use of all neutrons from the long ESS pulse, a sequence of sub-pulses will be selected that the wavelength band of subsequent sub-frames will overlap. Then the available wavelength band from 2 to 10Å will be completely covered thus avoiding gaps in Q. It is necessary, however, that the time sub-frames of these sub-pulses are well separated to avoid ambiguity in the wavelength (i.e. in Q) determination. The high-resolution choppers will be installed just behind Ch.1 and Ch.3 (see Fig. 4b).

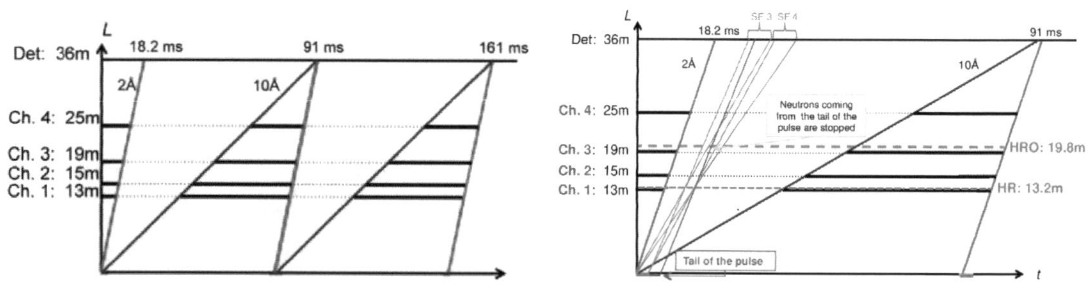

Fig. 4: Time-distance diagrams of the (a) low and (b) high resolution mode

Fig. 5: A double disc chopper allowing different openings by setting an offset angle between the discs.

Different resolutions require a variable opening of the pulse-shaping high-resolution chopper. Therefore we use a double disc chopper (see Fig. 5) that will allow different openings by setting an offset angle between the discs. For a 1% wavelength resolution mode, an offset angle of 45° will result in two openings of 10°. In the 3% and 5% modes, the offset angles will be set between -12.5° and +12.5°, leading to 4 windows and a variable opening from 0-25°. The frequency of the sub-pulses has to be an integer multiple of the frequency of the ESS source, otherwise the timing of the sub-pulses will continuously shift relative to the ESS pulses. The simulated time and wavelength intensity distributions for the 1, 3 and 5% resolution modes show a clear separation of the time sub-frames and an overlap of the wavelength sub-frames (see Fig. 6). It is important to note that in all three modes, the $\Delta\lambda/\lambda$ value is not constant because the pulse length, whether directly from the ESS or from the HM-chopper, is always constant, while the λ values change.

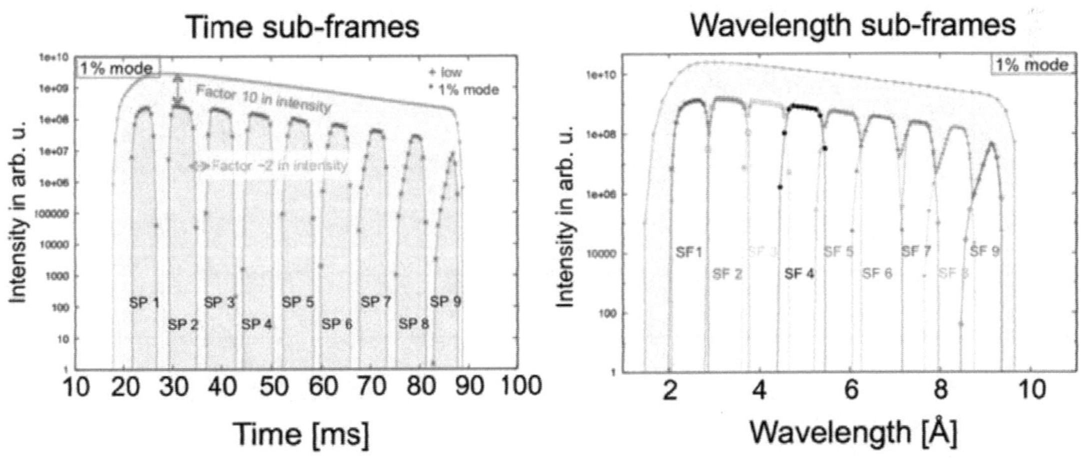

Fig. 6: The simulated time and wavelength intensity distributions for the 1% wavelength resolution modes. The low wavelength resolution curves ($\Delta\lambda/\lambda$=10%) are shown as envelops.

2.3. Polarization option:

In the polarized mode, the central 2m long piece of the guide (see Fig.2) will be replaced by another one equipped with a polarizing cavity which is built upon thin, 0.3mm thick, Si wafers coated with m=5 Fe/Si supermirrors working in transmission [2]. This solution leads to high values of neutron beam polarization (see Fig. 7) with small intensity losses over the whole wavelength band and allows practically for instant-switch between polarizing and non-polarizing operation modes without affecting the overall beam propagation. Even higher polarization of the neutron beam of more than 99% can be achieved using an optional ^{3}He neutron spin filter with a wide band neutron adiabatic RF-flipper, resulting in a drop of the beam intensity of about 25%. It should be noted that all sections after the polarizer will be coated by non-magnetic supermirrors.

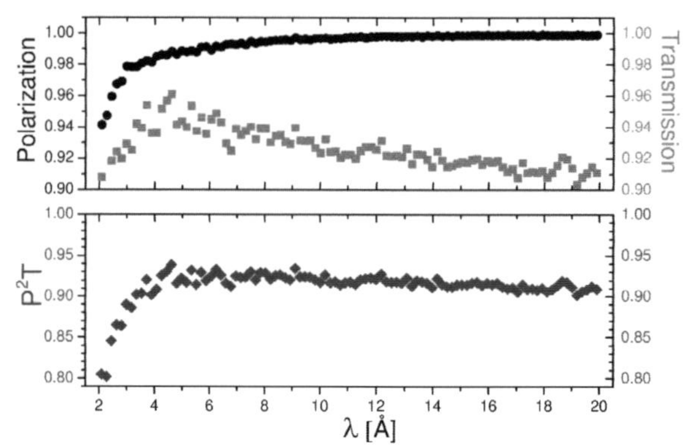

Fig. 7: Polarization P, transmission T of the selected spin component and the figure of merit P^2T of the polarizing cavity setup.

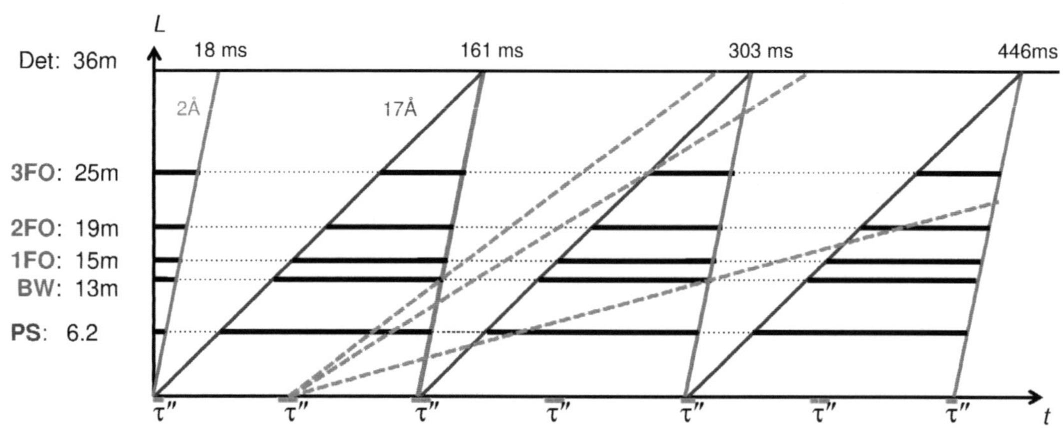

Fig. 8: Time-distance diagram for the kinetic mode of the 1 pulse skipping mode (7Hz), i.e. skipping every second pulse of the ESS source.

2.4. Kinetic mode:

For kinetic measurements it is desirable to cover a large Q-range for a single angular setting to allow measuring of the kinetic processes on time scales of one second and less. The design of VERITAS enables one to extend the Q range beyond the ratio of $Q_{max}/Q_{min} \approx 5$ in the basic chopper mode by skipping one or more pulses. This can be easily realized by placing one additional chopper (PS) at 6.2m distance from the source blocking the neutrons coming from the second (7Hz mode), third (4.7Hz) or fourth (3.5Hz) pulse as depicted in the time-distance diagram in Fig. 8 for the 1 pulse skipping mode. The corresponding choppers′ settings are listed in Table 2, and e.g. the 3 pulse skipping mode and enables one to measure a complete reflectivity curve in a Q range from 0.0075Å-1 (total reflectivity plateau) up to 0.12$Å^{-1}$.(~10^{-5} level depending on the sample) in one shot with a time resolution of 286ms as shown in Figure 9.

Mode	PS[°]	BW[°]	1FO[°]	2FO[°]	3FO[°]	f[Hz]	Time[ms]	Q-ratio
1 pulse	-66.2	-84.9	97.1	121.1	156.4	7	143	9
2 pulse	-40.9	-79.0	-90.3	-112.6	-145.3	4.7	212	12
3 pulse	-39.4	-76.0	-86.9	-108.3	-139.7	3.5	286	16

Table 2. The different parameters of the chopper settings used in the pulse skipping mode.

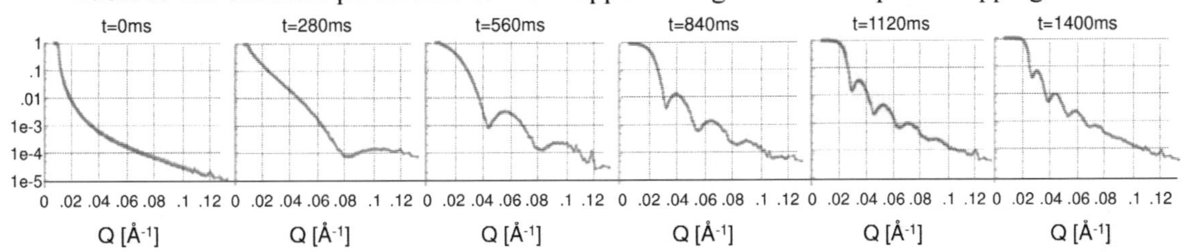

Fig. 9: Simulation of the reflectivity of a Ni thin film on a Si substrate growing in-situ with a growth rate of 28.5Å/s inside an MBE and recorded in a time interval of 280 ms on a 1cm^2 sample. The parameters are taken for the low resolution mode in the 3 pulse skipping mode.

3. Performance of the setup.

Taking the basic setup, the intensity-wavelength distribution is simulated for the proposed 10% wavelength resolution (see Fig. 10). The intensity scale is shown for a measurement time of one second and a beam collimation of 3mrad (2mrad Gaussian equivalent) at a spot size of 1x1cm^2. The integrated intensity over the full wavelength range amounts to 2.4x10^8 n/cm^2 for each single pulse of the ESS. For the integrated intensity per second this value has to be multiplied by the repetition rate of the ESS (14Hz) leading to 3.4x10^9 n/cm^2/s (see Fig. 10). The horizontal divergence profile is depicted in the inset of Fig. 10 showing a smooth distribution and demonstrating that the setup accepts the full divergence of the beam through the collimation. The slight wiggles in the profile are due to the polygonal structure used for the simulation of the curved guide. They will disappear by the use of continuously curved sections at the real instrument. In the GISANS mode the vertical focusing is disabled for a perfect collimation based on a fixed collimation length of 4m. The resolution can be adapted to the needed values in the vertical and horizontal direction. Fig. 11 shows the resulting intensity and the vertical divergence distribution (inset) at the sample position. A total flux of 1.0x10^8 n/cm^2/s at 3mrad collimation (2mrad in Gaussian approximation) in both direction is available with a clean vertical divergence profile. The integrated intensity over the full wavelength band will amount to $1.0 \cdot x10^8$ n/cm^2/s.

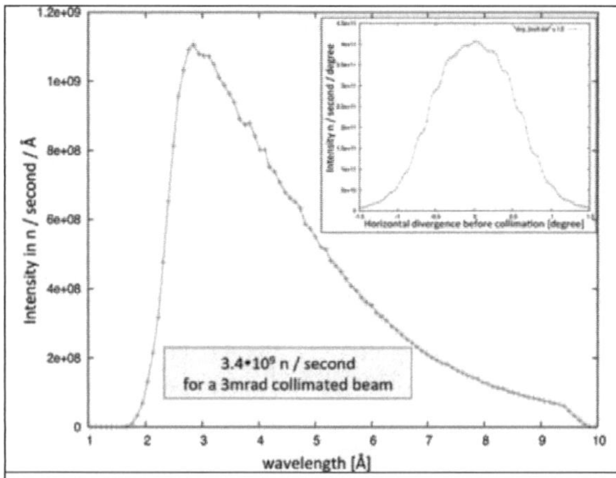

Fig. 10: Spectrum and divergence profile for the low resolution setup of the reflectometer in the lowest wavelength band from 2-10Å and a beam collimation of 3mrad (Boxcar)). The inset shows the horizontal divergence profile in front the collimation.

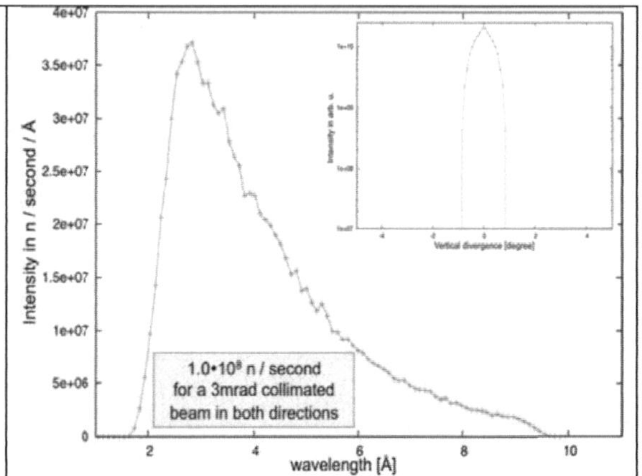

Fig. 11. Spectrum and divergence profile for GISANS setup in the low resolution setup of the reflectometer in the lowest wavelength band from 2-10Å and a beam collimation of 3mrad (Boxcar)). The inset shows the vertical divergence profile.

The suggested reflectometer will provide an extremely high flux, exceeding the flux which is currently achievable at the best reflectometers in the world. A 25 times gain is expected in comparison with D17 at the ILL [3]. The simulations above are supported by simple estimations: (i) the average flux at the ILL and ESS is about equal, the chopping of the beam will result in the losses of the useful neutron intensity because of the blocking of the neutron beam between the neutron pulses produced by the choppers; (ii) the opening time is defined by the ratio of the pulse width to their period that is approximately equal to 1/25; (iii) such pulse structure is naturally produced by the ESS, thus no losses related to the time structuring of the neutron beam will occur. Therefore, assuming that other beam parameters are similar, the expected gain is about 25.

4. Conclusions

The instrument concept presented above allows to design a very flexible vertical sample reflectometer for specular reflectivity and off-specular scattering as well GISANS investigations (in the pinhole geometry) at a modern long pulse source. An elliptically focusing neutron guide gives a gain of a factor 4 in intensity compared to a straight guide.

As the length of the instrument depends on the pulse structure of the source and the resolution of the instrument, VERITAS is designed for the ESS, delivering a maximum flux centered around 3Å with 10% wavelength resolution . It results in a stunning flux of $3.4x10^9$ n/cm^2/sec at the sample position for a 3mrad collimated beam (scattering plane) that is 25 times higher than one can achieve with an equivalent instrument at the ILL (e.g. D17). The wavelength resolution can be increased to e.g. 1%, 3% or 5% by using pulse-shaping choppers. Besides the reflectometry mode, VERITAS can be switched into the GISANS mode in seconds, preserving the gain of 25 over an equivalent instrument at the ILL. The huge neutron flux available at the sample position enables one to investigate systems where the sensitivity to a small amount of material or the resolution of small length scales around interfaces in thin film materials is required, something that is not feasible today.

The reflectometry mode as well as the GISANS mode can be further easily combined with a polarizer/analyzer system as well as with the kinetic mode of VERITAS. The need for the polarizer and analyzer system follow the increasing demand in the research field of magnetic nanoparticles (e.g.

assembled in thin films), magnetism, ferroelectricity and superconductivity at interfaces or novel topologically protected magnetic states (e.g. skyrmions) besides many other topics. In addition the polarizer/analyzer system can be used to measure precisely the incoherent background of soft matter samples, increasing the dynamic range of reflectivity and GISANS studies in this science area.

The VERITAS concept at the ESS (or adapted to any other long pulse source with similar strength or even steady reactors like PIK at the PNPI, Russia) allows one to push the limits in thin film science in all directions, making a huge step forward in the instrumentation, particularly for exploiting the full potential of the new spallation sources supporting the users in performing cutting edge science.

Acknowledgements

The authors would like to acknowledge the fruitful discussions with F. Cousin (LLB), R. Cubitt (ILL), Th. Brückel (FZJ), M. Lösche (CMU), F. Henrich (CMU) and Th. Gutberlet (HZB). We also would like to thank S. Manoshin (JINR, Dubna) for the support with VITESS and D.M. Rodriguez (ESS) for the initial VITESS simulations.

References

[1] http://eval.esss.lu.se/DocDB/0002/000274/015/TDR_online_ver_all.pdf
[2] Th Krist, C Pappas, A Teichert, C Fehr, D Clemens, E Steichele and F Mezei, Journal of Physics: ConferenceSeries 251 (2010) 012081
[3] R. Cubitt, G. Fragneto, Appl. Phys. A 74 [Suppl.], S329–S331 (2002)

Journal of Physics: Conference Series **711** (2016) 012010 doi:10.1088/1742-6596/711/1/012010

Polarisation Analysis Neutron Spectrometer, POLANO, at J-PARC - Concept and Magnetic Field Optimisation

K. Ohoyama[1,2]*, T. Yokoo[3,4], S. Itoh[3,4], M. Nanbu[2], K. Iwasa[5], M. Ohkawara[2], N. Kaneko[3,4], T. Ino[3,4], H. Hayashida[6,7], T. Oku[4,7], H. Kira[6], S. Tasaki[8], M. Takeda[6,7], H. Kimura[9] and T. J. Sato[9]

[1]Advanced Institute for Materials Research, Tohoku University, Sendai 980-8577, Japan
[2]Institute for Materials Research, Tohoku University, Sendai 980-8577, Japan
[3]Neutron Science Division, High Energy Accelerator Research Organization, Tsukuba, 305-0801, Japan
[4]J-PARC Center, Tokai, 319-1195, Japan
[5]Department of Physics, Tohoku University, Sendai 980-8578, Japan
[6]Comprehensive Research Organization for Science and Society, Tokai, 319-1106, Japan
[7]Japan Atomic Energy Agency, Tokai, 319-1195, Japan
[8]Department of Nuclear Engineering, Kyoto University, Kyoto, 615-8530, Japan
[9]Institute of Multidisciplinary Research for Advanced Materials, Tohoku University, Sendai, 980-8577, Japan

E-mail: `kenji.ohoyama.vs@vc.ibaraki.ac.jp`

Abstract. The status of the polarised neutron spectrometer constructed at the Japan Proton Accelerator Research Complex through a collaboration between Tohoku University and KEK will be reported. In particular, the optimisation of magnetic fields to minimise neutron-beam depolarisation using the finite element method will be discussed on the basis of several simulations using the finite element method..

1. Introduction

Tohoku University and the High Energy Accelerator Research Organization (KEK) are jointly constructing a polarisation analysis neutron spectrometer, POLANO, at beam line 23 (BL23: a decoupled H_2 moderator) in the second experimental hall of the Materials and Life Science Experimental Facility (MLF) at the Japan Proton Accelerator Research Complex (J-PARC) [1, 2]. POLANO will be the first dedicated polarised neutron spectrometer at MLF with polarisation analysis capability for spin dynamics investigation. The polarised neutron scattering technique is indispensable for novel materials science; it is an important and direct probe for observation of pure magnetic responses, nuclear-magnetic interference scattering, and spin chilarity states. As a scientific target, for instance, POLANO aims at observing characteristic spin excitations, so-called hour-glass type magnetic excitations in some of high-T_c superconductors [3]. It is thought that understandings of the characteristic magnetic excitations

* Present address: Graduate School of Science and Engineering, Ibaraki University, Nakanarusawa 4-12-1, Hitachi, 316-8511, Japan

are necessary to clarify the role of spin interactions in the superconductivity mechanism. For such investigations, the usable energy region around and above 50 meV is necessary. Another scientific target is electronic and magnetic multipole physics. For $4f$-electron systems, multipolar degrees of freedom, such as in the case of quadrupolar and octupolar moments, are the main topics of interest. Although inter-multipolar interactions are thought to be responsible for determining properties of $4f$ systems, the current understanding of this subject is quite limited because the relevant dynamics are often hidden by magnetic excitations and lattice dynamics. Thus, observations of multipolar excitations are a research target of POLANO. A characteristic of scattering by octupolar moments is Q dependence; the octupolar form factor has a maximum at a higher Q region ($Q{\sim}8\,\text{Å}^{-1}$), while magnetic form factors decrease with increasing Q [4]. Thus, POLANO will be used for observation in the high-Q ($Q \geq 8\,\text{Å}^{-1}$) region to detect octupolar excitations. To achieve this high-Q region, a scattering angle up to 130° will be covered with position-sensitive detectors with a diameter of 20 mm in the first stage, and an incident energy, $E_{\rm i}$, of over 80 meV will be used. We choose polarised neutron devices, as described later, because of such a demand for the neutron energy range.

The proposal of POLANO project was accepted by the J-PARC Center in 2011. The construction of the radiation shields was completed in 2013 and the vacuum chamber was installed on January 15th, 2015. The first beam is expected in 2016.

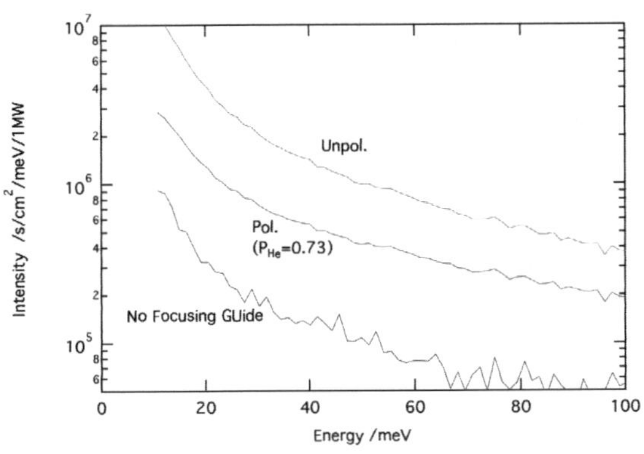

Figure 1. Expected white beam flux at the sample position calculated using McStas for $4Q_{\rm c}$ focusing guide tubes [2]. Red and blue lines indicate unpolarised and polarised beam flux; ^{3}He polarisation, $P_{\rm He} = 0.73$, ^{3}He gas pressure $= 0.203$ MPa, and the cell length $= 10$ cm. The black line indicates the flux without focusing guide tubes.

2. Flux and Polarisation Devices

To obtain high beam flux for polarised analysis experiments, focusing guide tubes with $4Q_{\rm c}$, supermirrors, where $Q_{\rm c}$ is the critical momentum transfer of Ni mirror, are installed; details of the focusing guide are reported in Ref. 2. Figure 1 shows the calculated results of the neutron beam flux at the sample position given by the focusing guide tubes, calculated by the Monte Carlo simulation using a program McStas. A cylindrical ^{3}He spin filter polarised by the on-beam spin exchange optical pumping (SEOP) technique will be installed at POLANO as a polariser. Because of the scientific targets briefly mentioned above, the SEOP-type ^{3}He spin filter is the best choice for the required energy region at the moment. The SEOP-type ^{3}He polariser is to be put on an elevator table so that the ^{3}He spin filter height can be controlled for optimisation and for operation in the unpolarised mode. Details of the SEOP technique and the design of POLANO polariser are being reported by Ino *et al.* [6].

For ^{3}He spin filters, the neutron polarisation and transmission are expressed as;

$$T_n = e^{-\sigma_a(\lambda)d}\cosh(P_{\rm He}\sigma_a(\lambda)d) \tag{1}$$

$$P_n = \tanh(P_{\rm He}\sigma_a(\lambda)d) \tag{2}$$

In this equation, $\sigma_a(\lambda)$ is the absorption cross section (cm^2) for the neutron wavelength, λ, while d is a parameter corresponding to the number of ^{3}He atoms in the spin filter cell. This is defined as $d = P \times l \times 2.414 \times 10^{20}$, where P and l are the ^{3}He gas pressure (MPa) at room temperature and the length of the cell (cm), respectively. The blue line in Fig.1 indicates the flux of polarised neutrons using a ^{3}He spin filter with $P_{He} = 0.73$, $P = 0.203$ MPa, and $l = 10$ cm; Note that $P_{He} = 0.73$ is the highest value in Japan [5]. A polarised beam flux of $\sim 3.6 \times 10^5$ /s/cm^2/meV can be expected to be obtained at $E = 60$ meV when the source power reaches 1 MW. On the other hand, Eqs.(1) and (2) indicate that higher neutron polarisation yields lower neutron transmission inevitably. Thus, for practical experiments, optimisation is important. An estimation of the optimisation of the parameters for ^{3}He spin filters was discussed elsewhere [1].

Polarisation analysers are an important technical issue to resolve the magnetic scattering component from the nuclear scattering one. For high-energy regions, spin filter analysers are a reasonable choice, but there are many technical difficulties that should be overcome in order to realise a spin filter analyser covering a large solid angle. Thus, as a practical choice, in the first stage of the project, a polarisation analyser composed of a set of Fe/Si supermirror benders will be installed; the bender blade is a $5.5Q_c$ mirror on the convex surface, and $1.5Q_c$ mirror on the concave surface. The analyser covers a 2θ range of approximately 16°, and will be installed on a table that rotates around the sample position in the vacuum chamber. The slit-width between the supermirror bender blades is 0.6 mm. The transmission of the supermirror analyser depends on the angular divergency of the incident neutron at the analyser entrance and the energy-dependent reflection angle on the supermirror surface. Figure 2 shows the energy dependence of calculated transmission which was defined as the ratio between the intensity at the entrance and exit positions of one supermirror slit set. The calculations were done under the planned POLANO configuration including the focusing guide tubes, and a standard vanadium cylinder (outer diameter: 25 mm, inner diameter: 24 mm, height: 40 mm) is set at the sample position; for the calculations, the McStas component, Pol_Bender, in which the mirrors were set to parallel, was used. As shown in Fig. 2, a transmission of ~ 0.18 can be expected at 20 meV, which is basically consistent with the data from the supermirror manufacturer.

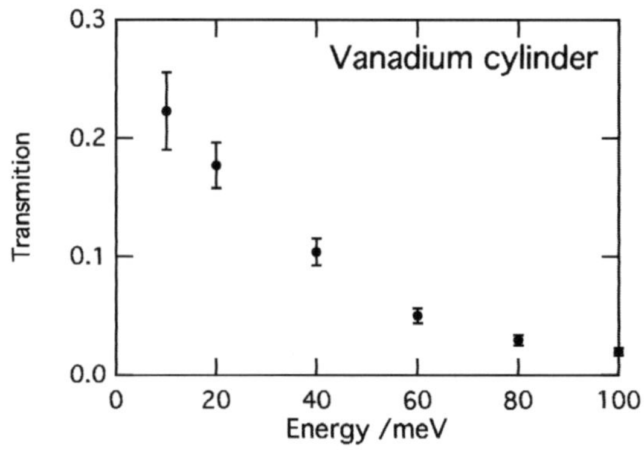

Figure 2. Transmission of one slit set only with a gap of 0.6 mm, which is defined as the ratio between the intensity at the entrance and exit positions of the slit under the POLANO configuration including the focusing guide tubes.

We have confirmed the required angular accuracy of the analyser. Figure 3 shows the integrated intensity for $E = 20$ meV at a detector from the standard vanadium cylinder. The horizontal axis is the angle of the analyser optical axis. Figure 3 means that an angular accuracy of approximately $\pm 0.2°$ is required in order to set the optical axis of the analyser to the maximum position. This condition is fulfilled by adopting the analyser table around the sample position that allows us to adjust the optical direction of the mirrors with an accuracy of 0.01° or less,.

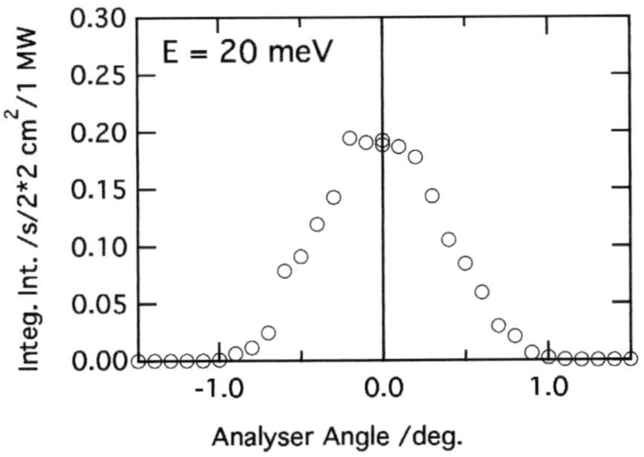

Figure 3. Integrated intensity for $E = 20\,\mathrm{meV}$ at a detector with a pixel of $2 \times 2\,\mathrm{cm}^2$ from a standard vanadium pipe. The horizontal axis is the analyser angle.

3. Magnetic Field Devices

Needless to say, the careful control of the magnetic fields around the beam is quite important for polarised neutron experiments. In this section, some results of estimation of magnetic fields are given. Firstly, we consider the magnetic field environment around POLANO. Only the minimum amount of iron is used in the radiation shield and vacuum chamber, but some iron materials are still present around POLANO; for instance, there are carbon steel nets under the floor of the experimental hall, and there is iron as part of the Fermi chopper pit, and as part of the radiation shields. Note that a side wall of the chopper pit which is nearest to the sample position is made of stainless steel to minimise the magnetisation effect. The effect of magnetisation of such iron materials on the polarised neutron beam should be estimated. Thus, as a worst case scenario, we considered the effects of 7 T magnetic fields generated by a MLF superconducting magnet on the iron materials using a finite element method program, ANSYS Maxwell.

Figure 4 shows the calculation model incorporating the structures containing iron materials in POLANO. The inset indicates the magnetic devices on the beam path, specifically: a solenoid coil for the SEOP-type ^{3}He spin filter, magnetic guides, a Helmholtz coil for samples, and a magnetic housing for the supermirror analyser. Figure 5 shows the magnitude of total magnetic fields along the beam path under the planned standard alignment of magnetic devices. The red line indicates the initial condition of POLANO calculated before application of the magnetic field of 7 T. After the calculation of the initial condition, the magnetic field of 7 T was applied at the sample position by a superconducting magnet, so that the iron around POLANO was magnetised by leaked magnetic fields from the superconducting magnet. The blue line indicates that calculated magnetic fields following application of the magnetic field of 7 T, which includes magnetic fields from the magnetised materials. Note that the 7T superconducting magnet itself was not included in the calculations. The difference between the data before and after application of the magnetic field is less than 1.5 G, which is much smaller than the typical guide fields ($\sim$20 G) along the beam path. This means that magnetisation effects of the iron parts around POLANO are negligible even though the iron is magnetised by the 7 T superconducting magnet. The nearest iron part to the sample position is a wall of a radiation shield, which is at a distance of approximately 140 cm or more. This indicates that the distance of $\sim$140 cm is sufficiently large to prevent serious magnetisation effects caused by leaked magnetic fields from a 7 T superconducting magnet.

Secondly, we optimised the magnetic-field devices to minimise the depolarisation between the ^{3}He spin filter and the analyser. When neutrons with finite velocity enter in a magnetic field, the neutron spin is rotated due to Larmor precession. The final spin wave function, ψ_f, due to

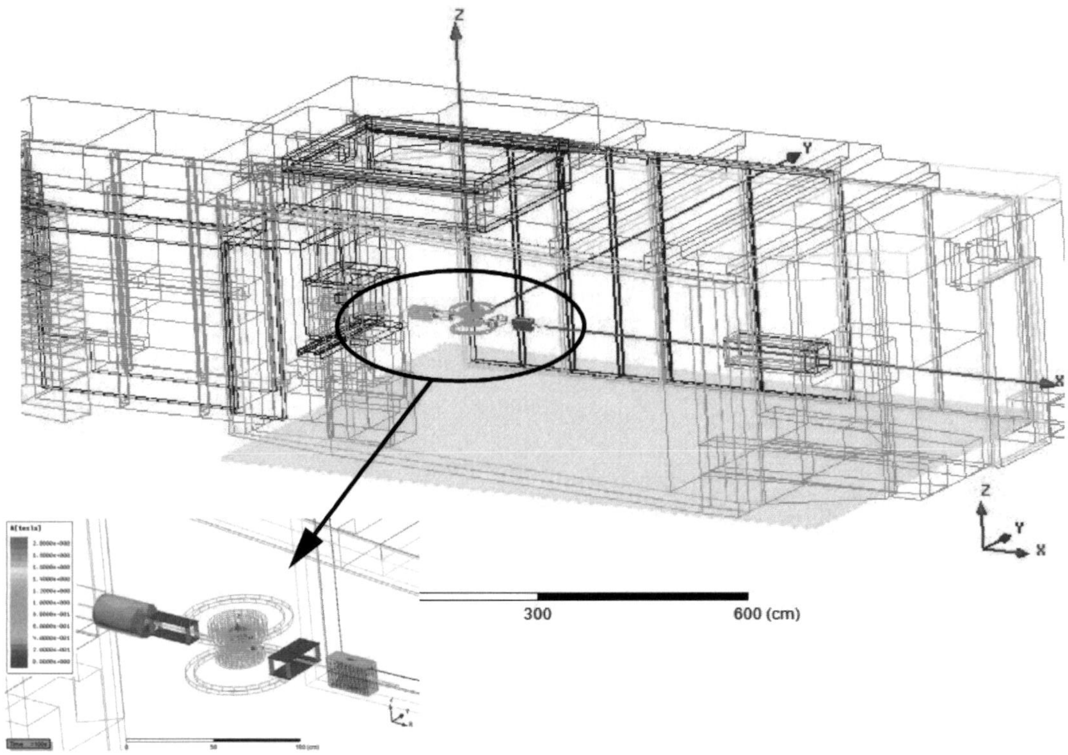

Figure 4. Model of estimation of magnetic fields, indicating iron parts around POLANO. The iron in radiation shields, the floor of the experimental hall which includes iron steel net 10 cm below the floor surface, some of the chopper pit walls, etc are included. Standard magnetic devices such as a solenoid coil for the SEOP-type ^{3}He spin filter, magnetic guides, the Helmholtz coil for samples, and a magnetic housing for the supermirror analyser,as shown in the inset, are also included in the calculations. The 7 T superconducting magnet is set at the sample position and generates a magnetic field of 7 T.

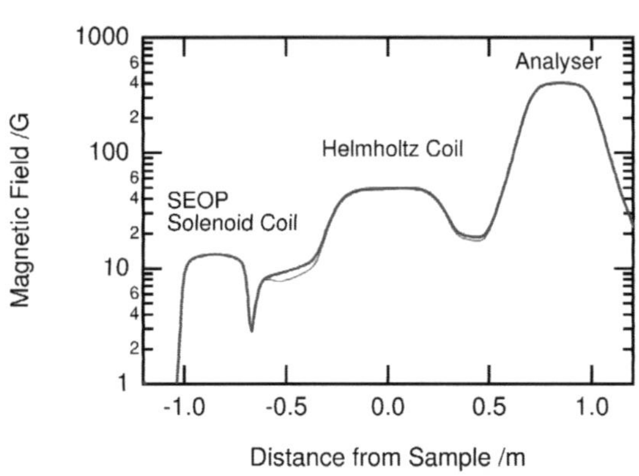

Figure 5. Effect on the beam path area from iron magnetised by a 7 T superconducting magnet calculated by ANSYS Maxwell. The red and blue lines indicate the magnetic fields provided by the magnetic devices and the iron parts around POLANO before and after application of 7 T magnetic field by the superconducting magnet.

Larmor precession can be represented using the initial spin wave function, ψ_i, as

$$\psi_f = \exp\left(-i\frac{\boldsymbol{S}\cdot\boldsymbol{n}}{\hbar}\frac{2\omega l}{v}\right)\psi_i \tag{3}$$

where ω, l, and v are Larmor frequency, the length of the flight path under the magnetic field, and neutron velocity, respectively, and $\boldsymbol{S}$ is the neutron spin. $\boldsymbol{n}$ is the unit vector of the magnetic field, $\boldsymbol{B}$, at l. The average value of each spin component in ψ_f is obtained as $\langle S_\alpha \rangle = \hbar/2\ \psi_f^* \sigma_\alpha \psi_f$ ($\alpha = x, y, z$), where σ_α is the Pauli matrix. From the average spin components, the neutron polarisation, P_n, of the final state can be defined as

$$P_n = \frac{\langle S_x \rangle B_x + \langle S_y \rangle B_y + \langle S_z \rangle B_z}{|\boldsymbol{S}||\boldsymbol{B}|} \tag{4}$$

Thus, when ψ_i and the magnetic field distribution with an interval of l are provided, the value of P_n at each point can be estimated. For these calculations, the z-axis is parallel to the beam direction, the y-axis is in the vertical direction, and the x-axis is perpendicular to the y- and z-axes. Since the magnetic field at the ^{3}He spin filter is parallel to the beam direction, the initial neutron spin is parallel to the z-axis. Note that l must be sufficiently short to allow the magnetic field to be treated as a constant. In the calculations below, we calculated P_n with $l = 1$ mm condition because we confirmed that the typical results of magnetic field distribution with $l = 1$ mm and 0.1 mm gave the same P_n at each position.

We calculated the magnetic field distribution along the beam path within the area of x, $y = \pm 3$ cm with magnetic devices of a solenoid coil for the SEOP-type ^{3}He spin filter (current: 2 A), magnetic guides, a Helmholtz coil for samples, and a magnetic housing for the supermirror analyser. The SEOP-type ^{3}He spin filter is the same one as in Ref.6. Figures 6(a) and (c) show the total magnitude (B) of the magnetic field at the beam centre (a) and the $y = $-3 cm (c) position below the beam centre. These calculations were done using ANSYS in SR16000 supercomputing resources at the Center for Computational Materials Science of the Institute for Materials Research, Tohoku University. From the magnetic field distributions in Fig. 6, $\langle S_\alpha \rangle$ and P_n can be obtained from Eqs. (3) and (4). We calculated P_n in the energy region between 10 and 100 meV in the beam area of -3 cm $\leq x, y \leq 3$ cm. As typical results, Figs. 6(b) and (d) indicate the change in P_n for $E = 80$ meV along the beam path at $x = 0$ cm. At the beam centre ($y = 0$ cm) (Fig. 6(b)), P_n is almost constant except the slight depolarisation at $z \sim 0.2$ m (blue triangle), and the final P_n at the analyser ($z \sim 1.7$ m) is 0.96, which is satisfactory for practical experiments. On the other hand, for $y = $-3 cm, P_n decreases drastically after the exit of the SEOP solenoid coil at $z \sim 0.2$ m. The obvious depolarisation is caused by the suppressed magnetic fields (blue triangle), where the magnetic fields generated by the SEOP solenoid coil and vertical guide magnet are almost cancelled; the suppressed magnetic field is at $\sim$6 G, and the final P_n is $\sim$0.14. To prevent the depolarisation found in Figs. 6(b) and(d), a additional solenoid guide coil with a length of 3 cm at $z \sim 0.27$ cm is effective to cover the bottom of the magnetic field. The dashed red lines in Figs. 6(b) and (d) shows P_n when the 3 cm coil is used (current: 1.5 A), which increases the bottom of B around z=0.2 m above 20 G; the depolarisation in Fig. 6(d) is drastically improved, and the final P_n is almost 0.99. Note that the slight depolarisation in Fig. 6(b) around $z = 0.2$ m also disappears. Based on systematic calculations of depolarisation, we found empirical criteria of the magnetic field that can prevent depolarisation; when B is 20 G or larger, depolarisation can be neglected even in the energy range between 10 and 100 meV in the alignment of POLANO.

On the other hand, there is a possibility that additional coils near the SEOP solenoid coil deteriorate the homogeneity of the magnetic field in the SEOP solenoid coil, which affects the relaxation time of the ^{3}He spin, and thus, P_{He} [6]. We calculated the effect of the 3 cm coil and other devices to the homogeneity of the magnetic field in the SEOP solenoid coil based on Ref.6. The required homogeneity of the magnetic field at the ^{3}He spin filter is expressed as

$$\frac{\mathrm{d}B_r}{\mathrm{d}r}\frac{1}{B_z} < 0.0008 \tag{5}$$

Journal of Physics: Conference Series **711** (2016) 012010 doi:10.1088/1742-6596/711/1/012010

where B_r and B_z are the radial and longitudinal direction components of the magnetic field in the SEOP solenoid coil, respectively, at the radial position r (cm), and at the longitudinal direction position, z (cm). We confirmed that the homogeneity of the magnetic field in the SEOP solenoid coil in the alignment with the 3 cm coil satisfies the requirement of Eq. (5) in the area of the ^{3}He spin filter, meaning that the alignment of the magnetic devices including the 3 cm coil found in the present calculations does not affect the P_{He}.

For the calculations mentioned above, the magnetic field applied by the Helmholtz coil is vertical at the sample position. We also confirmed that the same alignment of magnetic devices maintains P_n larger than 0.95 in the energy range between 10 and 100 meV, even for application of a horizontal magnetic fields at the sample position. Thus, the alignment found in the present work makes both the $P_n//Q$ and $P_n \perp Q$ experimental conditions possible; these conditions are indispensable for the separation of pure magnetic and nuclear scattering components.

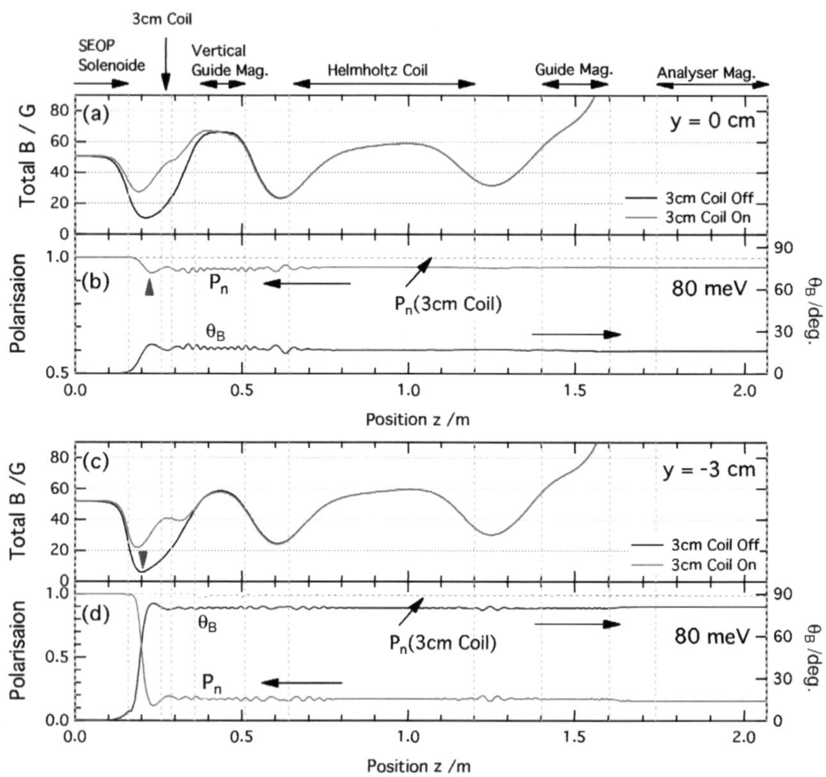

Figure 6. Results of estimation of depolarisation along the beam path for $E = 80$ meV. The horizontal axis is the distance along the beam direction from the centre of the SEOP solenoid coil. (a) and (c) show the total magnetic field at each point at the beam centre ($x = 0$ cm, $y = 0$ cm) and the $x = 0$ cm, $y = -3$ cm position. Note that the Helmholtz coil for the sample apples a vertical magnetic field at the sample position in this calculation. (b) and (d) show the neutron polarisation expressed according to Eq. 4, and the angle, θ_{B}, between the spin and magnetic field at each point. The vertical lines indicate the edges of each magnetic device. Note that the scales of the vertical axis of (b) and (d) are different. The red solid and dashed lines in (b) and (d) indicate P_n when the 3 cm coil at $z \sim 0.27$ cm is not used and is used (current: 1.5 A).

4. Summary

The first dedicated polarised neutron spectrometer, POLANO, at J-PARC is under construction through a collaboration between Tohoku University and KEK. The first beam is expected in the autumn of 2015. To obtain higher neutron flux, the $4Q_c$ supermirror guide tubes are installed; a polarised beam flux of $\sim 3.6 \times 10^5$ /s/cm^2/meV at $E = 60$ meV can be expected when the source power reaches 1 MW, with a cylindrical ^{3}He spin filter polarised by the SEOP technique. The transmission of the supermirror analyser is estimated to be 0.18 at $E = 20$ meV. We investigate the effect of magnetisation of iron around the beam line on the polarised neutron transportation. The iron in the radiation shields and the carbon steel net in the floor were taken into account, and magnetisation of these materials induced by a magnetic field of 7 T applied by a superconductor magnet was calculated. Using the finite element method, we confirmed that the effect of the magnetic fields from the magnetised iron on the polarised neutron beam is negligible. The alignment of magnetic devices on the beam path was optimised using a quantum physics treatment of spin transportation. From the calculations, we succeeded in finding an alignment of magnetic devices which gives satisfactory P_n in the energy range between 10 and 100 meV for both the $P_n // Q$ and $P_n \perp Q$ conditions.

Acknowledgments

The calculations in Fig. 6 has been performed under the inter-university cooperative research program of the Center for Computational Materials Science, Institute for Materials Research, Tohoku University (Proposal No.: 13S0416, 14S0413). This work was partly supported by the S-type project of KEK (2009S09, 2014S09) and a Grant-in-Aid for Challenging Exploratory Research (24656004), Scientific Research (A)(23244068), (S)(23224009).

References

[1] K. Ohoyama, T. Yokoo, S. Itoh, J. Suzuki, K. Iwasa, T.J. Sato, H. Kira, Y. Sakaguchi, T. Ino, T. Oku, K. Tomiyasu, M. Matsuura, H. Hiraka, M.Fujita, H. Kimura, T. Sato, J. Suzuki, H.M. Shimizu, T. Arima, M. Takeda, K. Kaneko, M. Hino, S. Muto, H. Nojiri, C.H. Lee, J.G. Park, and S. Choi, J. Phys. Soc. Jpn. **82** (2013) SA035.

[2] T. Yokoo, K. Ohoyama, S. Itoh, J. Suzuki, K. Iwasa, T. J. Sato, H. Kimura, N. Kaneko, and M. Ohkawara, J. Phys.: Conf. Ser. **502** (2014) 012046.

[3] For instance, J. M. Tranquada, H. Woo, T. G. Perring, H. Goka, G. D. Gu, G. Xu, M. Fujita, and K. Yamada, Nature **429** (2004) 534.

[4] R. Shiina, O. Sakai, and H. Shiba, J. Phys. Soc. Jpn., **76** (2007) 094702.

[5] H. Kira, Y. Sakaguchi, T. Oku, J. Suzuki, M. Nakamura, M. Arai, K. Kakurai, Y. Endo, Y. Arimoto, T. Ino, H.M. Shimizu, T. Kamiyama, K. Tsutsumi, K. Ohoyama, H. Hiraka, K. Yamada, and L.J. Chang, Physica B **406** (2011) 2433.

[6] T. Ino, K. Ohoyama, T. Yokoo, S. Itoh, M. Ohkawara, H. Kira, H. Hayashida, K. Sakai, K. Hiroi, T. Oku, K. Kakurai, and L. J. Chang, Proceedings of Polarised Neutrons for Condensed-Matter Investigations 2014

Conceptual design of a polarized ^{3}He neutron spin filter for polarized neutron spectrometer POLANO at J-PARC

T Ino[1,2], K Ohoyama[3,4], T Yokoo[1,2], S Itoh[1,2], M Ohkawara[4], H Kira[5], H Hayashida[5], K Sakai[6,2], K Hiroi[6,2], T Oku[6,2], K Kakurai[6], and L J Chang[7]

[1] IMSS, KEK, Tsukuba, Ibaraki 305-0801, Japan
[2] J-PARC Center, Tokai, Ibaraki 319-1195, Japan
[3] WPI-AIMR, Tohoku University, Sendai, Miyagi 980-8577, Japan
[4] IMR, Tohoku University, Sendai, Miyagi 980-8577, Japan
[5] CROSS, Tokai, Ibaraki 319-1106, Japan
[6] JAEA, Tokai, Ibaraki 319-1195, Japan
[7] Physics Department, National Cheng Kung University, Tainan 70101, Taiwan

takashi.ino@kek.jp

Abstract. A ^{3}He neutron spin filter (NSF) has been designed for a new polarized neutron chopper spectrometer called the Polarization Analysis Neutron Spectrometer with Correlation Method (POLANO) at the Materials and Life Science Experimental Facility of the Japan Proton Accelerator Research Complex. It is designed to fit in a limited space on the spectrometer as an initial neutron beam polarizer and is polarized *in situ* by spin exchange optical pumping. This will be the first generation ^{3}He NSF on POLANO, and a polarized neutron beam up to 100 meV with a diameter of 50 mm will be available for research on magnetism, hydrogen materials, and strongly correlated electron systems.

1. Introduction

A Polarization Analysis Neutron Spectrometer with Correlation Method (POLANO) is in the final stage of construction at the Materials and Life Science Experimental Facility (MLF) of the Japan Proton Accelerator Research Complex (J-PARC). POLANO is a new inelastic neutron spectrometer, focusing on quasi-elastic and inelastic scattering techniques, for comprehensive material research with polarized neutrons [1]. Some of the basic instrumental parameters of POLANO are listed in Table 1.

Incident neutron energy	up to 200 meV
Energy resolution $(\Delta E/E_i)$	3 ~ 5% at elastic position
Momentum resolution $(\Delta Q/k_i)$	1 ~ 2%
Sample size	20 mm × 20 mm
Detector coverage	−30°~130° (horizontal)
	−10°~10° (vertical)

Table 1 Some basic instrumental parameters of POLANO.

A neutron polarizer as well as a neutron-spin analyzer is a key device for the successful operation of POLANO. A ^{3}He neutron spin filter (NSF) will be installed on the spectrometer as an initial neutron beam polarizer, and polarizing supermirrors as neutron-spin analyzers. The ^{3}He NSF will be continuously operated (*in situ* polarization) by spin exchange optical pumping (SEOP) [2] to provide a neutron beam with a constant polarization [3,4]. It will be placed at a distance of 16.5 m from the neutron source, between the collimators and the sample vacuum chamber, as illustrated in Fig. 1. The available space for the ^{3}He NSF is limited, roughly 60 cm by 80 cm with the shorter side lying along the neutron beam axis. The other requirements for the first generation ^{3}He NSF on POLANO are:

- a neutron polarization $P_n > 90\%$ at 100 meV
- a neutron beam diameter > 50 mm.

A converging guide mirror has been installed upstream, and the beam size at the ^{3}He NSF position will be 50 mm × 50 mm to 80 mm × 80 mm depending on the neutron energy. Hence in the next generation ^{3}He NSF, the ^{3}He cell diameter will be increased to ~100 mm with the attainable neutron energies being as high as 200 meV.

Polarization of the ^{3}He nuclei by SEOP requires a glass cell of ^{3}He gas with a drop of alkali metal, a homogeneous magnetic field to hold the polarized ^{3}He spins, heaters to vaporize the alkali metal in a ^{3}He cell, and a high-power laser for optical pumping. An adiabatic fast passage nuclear magnetic resonance (AFP NMR) system is additionally required for POLANO to flip the ^{3}He spins for reversing the neutron spin direction. In addition, real-time whole-system monitoring with a safety interlock is essential for long-term operation of an *in situ* ^{3}He NSF. These components are discussed in detail in the following sections.

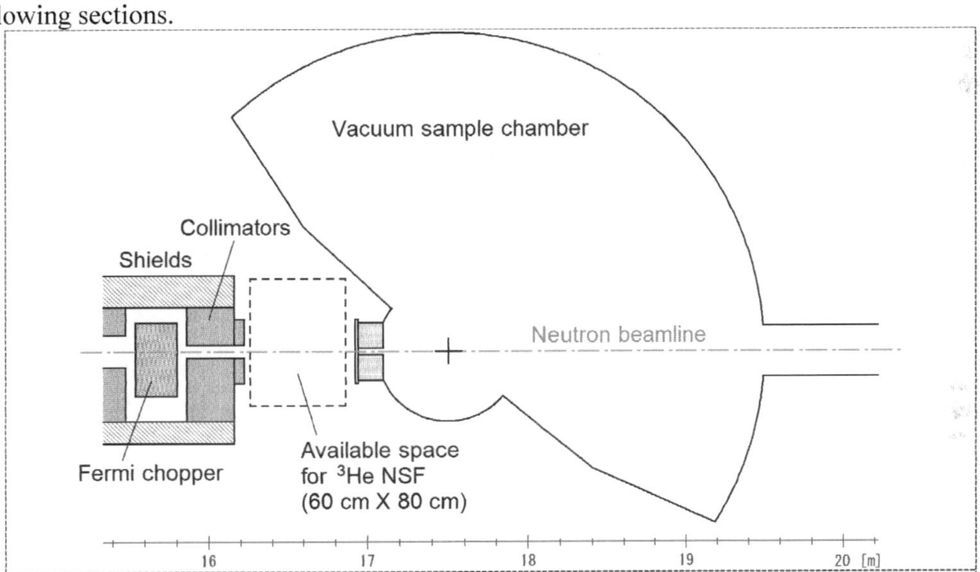

Figure 1. Schematic of a part of POLANO. An *in situ* polarized ^{3}He neutron spin filter will be located between the collimators and the sample vacuum chamber, 16.5 m from the neutron source. The available space for the ^{3}He NSF is about 60 cm by 80 cm.

2. ^{3}He cells

Cylindrical ^{3}He cells with diameters of 60 mm, that fully cover the required beam diameter of 50 mm, are to be used. They are made of aluminosilicate glass GE-180, which is now commonly used for ^{3}He NSFs [4]. The ^{3}He gas thickness[1] d_{He} necessary to achieve a certain neutron beam polarization P_n can be formulated as

[1] We refer to the product of the ^{3}He gas density at 0 °C and the length along the neutron beam direction as the ^{3}He gas thickness.

$$P_n = \tanh(P_{He}\sigma d_{He}) \quad \text{or} \quad d_{He} = \frac{\tan^{-1} P_n}{\sigma P_{He}}, \tag{1}$$

where P_{He} is the ^{3}He polarization, and σ is the spin-averaged neutron absorption cross-section of ^{3}He. Figure 2 shows the relations between the ^{3}He gas thickness and the neutron energy for $P_n = 0.8$, 0.9, and 0.95 with $P_{He} = 0.7$. To achieve a neutron polarization of 90% at 100 meV with a ^{3}He polarization of 70%, a ^{3}He thickness of ~30 atm·cm is required, which may be realized with a cell 10 cm long filled with ^{3}He gas at 3 atm (0 °C) or 3 amagat[2]. Shorter and longer cells could be used with higher and lower ^{3}He densities, respectively. However, both shorter and longer cells have drawbacks; longer cells require larger space and shorter cells with higher gas densities (pressures) portend structural difficulties in the fabrication of ^{3}He cells and limit the ^{3}He polarization owing to the magnetic-dipole interaction of ^{3}He. The ^{3}He spin relaxation by the magnetic-dipolar interaction between the ^{3}He nuclear spins is proportional to the ^{3}He gas density [5]. At 23 °C, the relaxation rate becomes

$$\frac{1}{T_{dipole}} = \frac{\eta_{He}}{744} \quad \text{[/hour]}, \tag{2}$$

where T_{dipole} is the relaxation time by the magnetic-dipole interaction and η_{He} is the ^{3}He density in amagat. In a ^{3}He cell filled at 3 amagat, the magnetic-dipole relaxation time drops to ~250 h at room temperature, and higher gas densities further shorten the relaxation time.

The windows of a ^{3}He cell should be as thin as possible to minimize the background neutrons scattered by the glass windows, but thick enough to withstand the gas pressure. Studies to develop ^{3}He cells that can sustain higher pressures are underway [6].

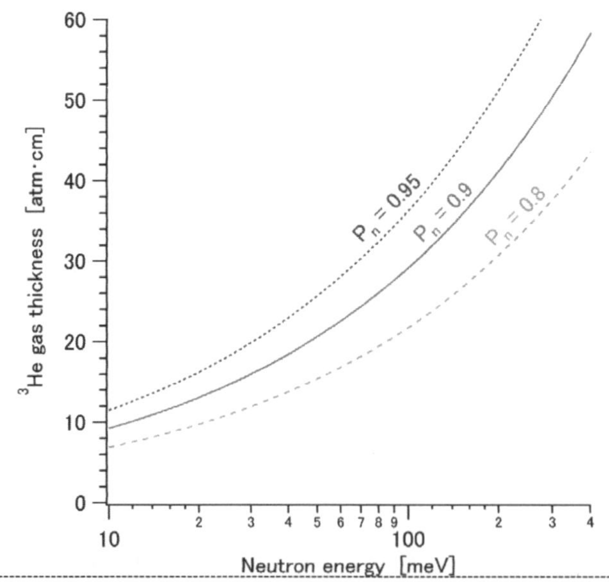

Figure 2. The required ^{3}He gas thickness is plotted against the neutron energy for neutron polarizations of 0.8, 0.9, and 0.95. The ^{3}He polarization is assumed to be 70%.

[2] An amagat is a unit of gas density corresponding to 1 atm at 0 °C.

3. Magnetically shielded solenoid

A homogeneous magnetic field is essential for polarized ^{3}He nuclei since the ^{3}He nuclear spin is quite sensitive to field gradients perpendicular to the principal axis of the magnetic field, and the polarization easily degrades in the presence of small field gradients [7]. The ^{3}He spin relaxation by radial (transverse) field gradients T_{mag} can be written as

$$\frac{1}{T_{mag}} = \frac{6400}{\eta_{He}}\left(\frac{1}{B_0}\cdot\frac{\partial B_r}{\partial r}\right)^2 \quad \text{[/hour]}. \tag{3}$$

Here, $(\partial B_r/\partial r)/B_0$ is the radial field gradient in [/cm] normalized by the primary field B_0. At $\eta_{He} = 3$ amagat, a radial field gradient of 0.001 /cm corresponds to $T_{mag} \approx 500$ h, which is twice as long as that by magnetic-dipole interaction, and therefore we have designed a magnetically shielded solenoid that yields radial field gradients less than 0.001 /cm in the cell location.

Figure 3 illustrates the structure of a magnetically shielded solenoid. The diameter and length of the solenoid are 20 cm and 30 cm, respectively. Two additional coils are wound around the solenoid at both ends; the solenoid and the two correction coils, for which insulated wires with a diameter of 1.0 mm are used, are connected in series so that they can be driven by a single power supply. Permalloy sheets wrap the coils for magnetic shielding with two openings at both the ends of the solenoid for the neutron and laser beams. The magnetic field distribution around the center of the solenoid and the radial field gradients have been calculated for a coil current of 1 A, as shown in Fig. 4. From the calculation, the transverse field gradients are much smaller than 0.001 /cm and satisfy our requirement in the volume of the cylindrical cell with a diameter of 60 mm and a length of 100 mm.

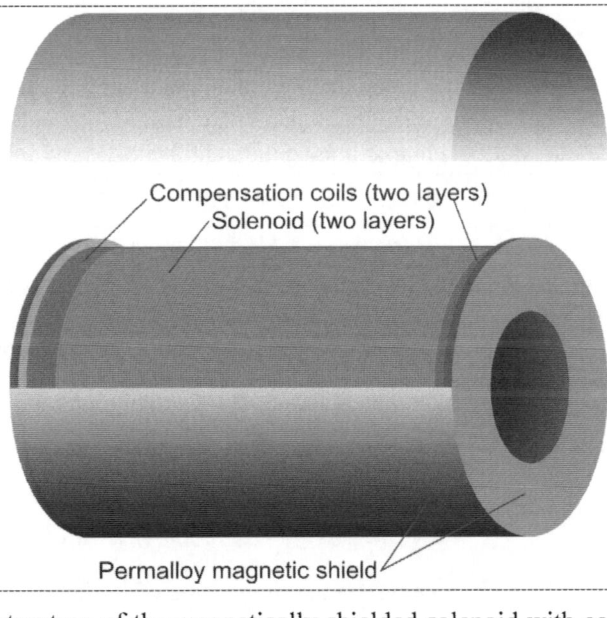

Figure 3. The structure of the magnetically shielded solenoid with correction coils for the ^{3}He NSF on POLANO. An insulated wire with a diameter of 1 mm is used for the solenoid (300 turns × 2 layers) and two correction coils (11 turns × 2 layers each).

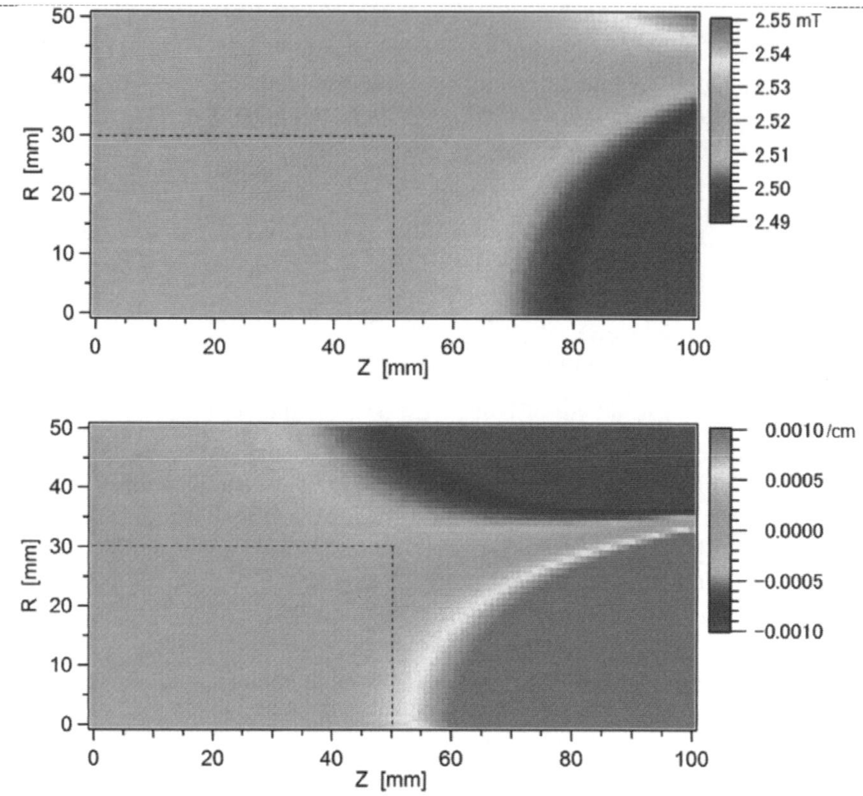

Figure 4. The calculated magnetic field distribution (top) and the radial field gradients (bottom) inside the magnetically shielded solenoid for a current of 1 A. Both are plotted on the rz plane in the cylindrical coordinate system with the origin at the center of the solenoid. The location of the cylindrical ^{3}He cell of diameter 60 mm and length 100 mm is superimposed as dashed lines.

4. Oven with AFP NMR

During the optical pumping of a ^{3}He cell by SEOP, the temperature in the oven is kept at around 200 °C to evaporate the alkali metal. An aluminum oven is used for the ^{3}He NSF on POLANO with AFP NMR to enable flipping the ^{3}He spins instantaneously [8]. A uniform radio-frequency (RF) magnetic field is essential to avoid polarization loss in AFP NMR. The AFP conditions can be written as

$$D\frac{\left|\vec{\nabla}B_z\right|^2}{B_1} << \frac{dB_z}{dt} << \gamma B_1^2 \quad \text{or} \quad D\frac{\left|\vec{\nabla}B_z\right|^2}{B_1} << \gamma\frac{d\omega}{dt} << \gamma B_1^2, \tag{4}$$

where ω is the RF magnetic field frequency, B_1 the magnitude of the RF field, $\vec{\nabla}B_z$ the gradient of the holding field B_z, γ the gyromagnetic ratio of ^{3}He, and D is the ^{3}He diffusion constant [9]. To fulfill the AFP conditions, a cosine-theta coil has been designed to produce a uniform RF magnetic field inside the oven (Fig. 5), and the calculated RF magnetic field is presented in Fig. 6. The variation of the RF field strength is less than +/-10% inside the ^{3}He cell location, and should be uniform enough to satisfy the AFP conditions. It is to be noted that the RF magnetic fields yielded by the coil wire leads outside the oven, extracted at both ends, do not affect the inside of the oven because the aluminum body of the oven is much thicker (5 mm) than the skin depth of aluminum at the RF frequency of AFP NMR[3] and shields RF interference from the outside.

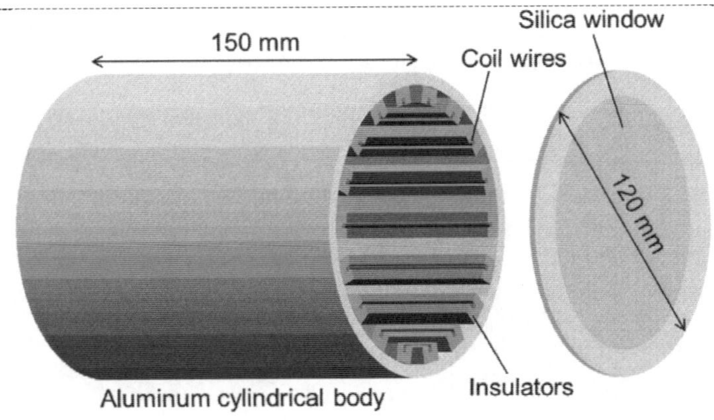

Figure 5. A sketch of the aluminum oven with a cosine-theta RF coil for AFP NMR. Wires are arranged at 20 degree intervals with a cosine-theta current distribution inside the oven.

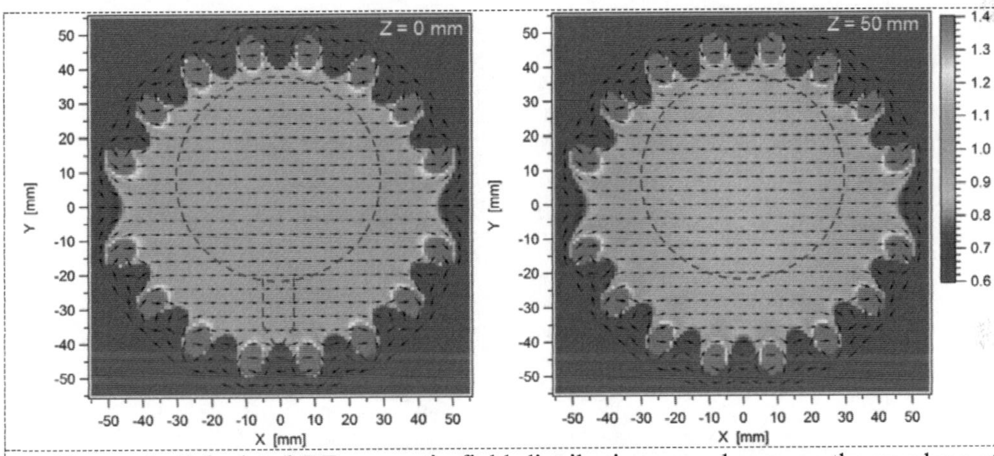

Figure 6. Calculated RF magnetic field distributions are shown on the xz plane at z = 0 mm and 50 mm inside the aluminum oven (Fig. 5). The relative field strengths normalized at the oven center and the field directions are plotted together with the ^{3}He cell (dashed line). Owing to the inlet pipe for the ^{3}He cell, the cylindrical body of the cell is off the center of the oven.

5. Oven heater

Hot air is commonly used to keep ^{3}He cells at elevated temperatures during optical pumping because conventional heating wires produce magnetic fields that interfere with the spin-holding magnetic field, causing additional depolarization. We, however, plan to use heating wires for the ^{3}He NSF on POLANO to keep the system compact and for safety reasons. Flexible sheet heaters, which are made of resistive strips sandwiched between thin polyimide films, are widely available. The strip patterns of the heater element can be customized, and we have studied these strip patterns to obtain ones that minimize the magnetic interference to the ^{3}He cell. Two patterns of the heater element are shown in

[3] The skin depth of aluminium is ~0.3 mm at 80 kHz, which corresponds to the Larmor frequency of ^{3}He in a static magnetic field of ~2.5 mT.

Fig. 7. The magnetic fields resulting from them have been calculated for a heater current of 1 A and are plotted in Fig. 8. Assuming the primary magnetic field is ~2.5 mT, the additional field gradients due to the presence of the sheet heater current will not exceed 0.0003 /cm in the ^{3}He cell position in either strip pattern. Note that the required heating capacity expected from preliminary tests of similar heaters is 100 W or less, much lower than that of a hot air heater used in Ref. 10.

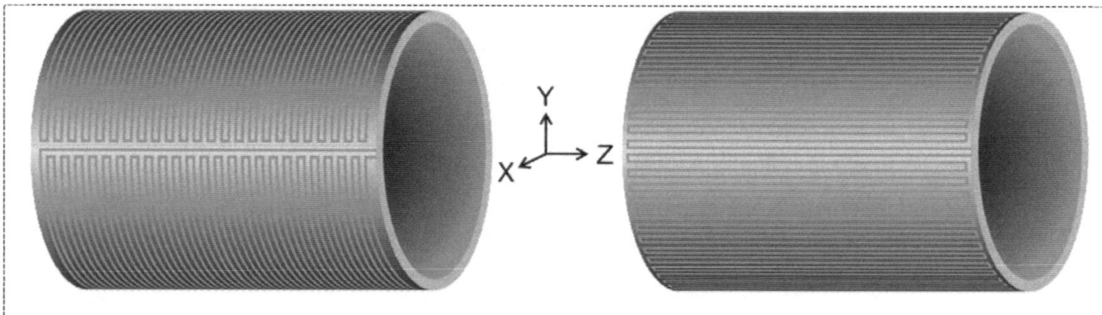

Figure 7. Flexible sheet heaters with the heater elements in radial (left) and axial (right) strip patterns are wrapped around the ovens. Each heating strip is 1 mm wide and is arranged at 3 mm intervals.

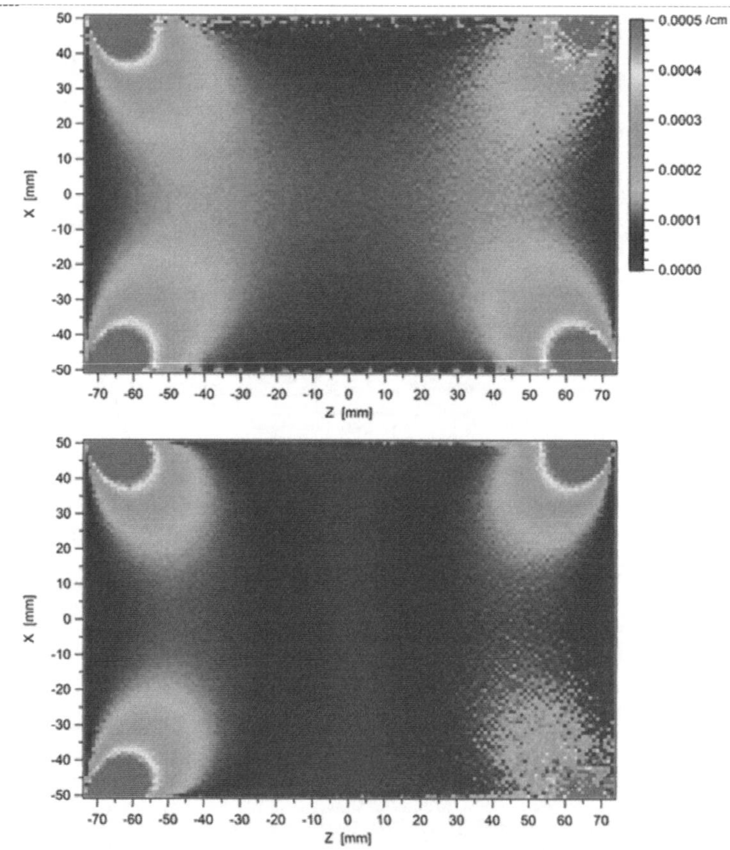

Figure 8. Magnetic field gradients due to the flexible sheet heaters calculated for a heater current of 1 A. The uniform spin holding magnetic field is assumed to be 2.5 mT along the z direction.

6. Frequency narrowed diode laser optics

With the advances in high-power semiconductor diode lasers, the ^{3}He polarization as well as the polarized gas volume has increased significantly. External cavity diode lasers have further improved the efficiency of optical pumping by reducing the linewidth of a diode laser [11,12].

We have tested and found a high-power laser diode array (LDA) with a chirped volume holographic grating (VHG) to be suitable for the ^{3}He NSF on POLANO owing to the performance and compactness. The grating constant of a chirped VHG varies with its translation position, and its emission wavelength can be finely tuned to the D_1 resonance of an alkali metal [12]. Figure 9 shows our measurements of emission spectra from an LDA with and without a chirped VHG. The full width at half maximum (FWHM) was successfully narrowed to less than 0.2 nm by the use of a chirped VHG. The FWHM is comparable to that obtained from our previous external cavity optics with a diffraction grating [13] but is achieved with smaller and lesser number of optical components.

Circularly polarized light for optical pumping has to be reversed in helicity according to the ^{3}He spin direction that can be flipped by AFP NMR. The helicity of the laser radiation can easily be reversed by rotating a quarter wave plate, thus converting the linearly polarized light from an LDA to circular. The quarter wave plate is mounted on a rotational holder, and the angle of rotation is remotely controlled synchronously with AFP spin flipping of the polarized ^{3}He.

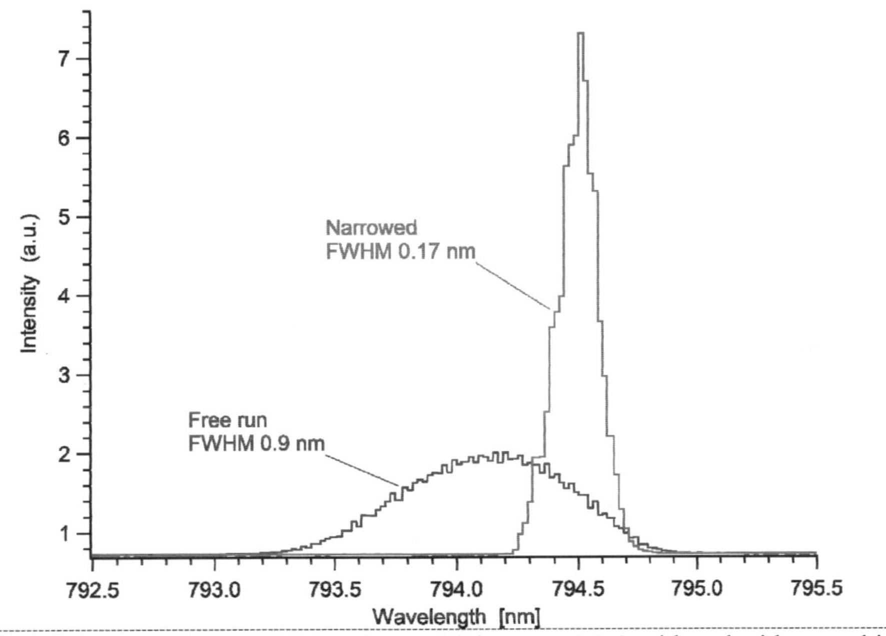

Figure 9. Laser emission spectra from a high-power LDA with and without a chirped VHG. Output power from the LDA was ~85 W.

7. Summary

We have designed an *in situ* operable ^{3}He NSF for initial neutron beam polarization on POLANO. Neutron polarizations > 0.9 and energies up to 100 meV may be achieved for a neutron beam diameter > 50 mm. A schematic of the ^{3}He NSF of dimensions 50 cm × 50 cm × 30 cm (length × width × height) is shown in Fig. 10. The laser illuminates a ^{3}He cell from one side in the drawing, but the laser optics will be modified to illuminate both sides of a ^{3}He cell with two lasers to improve the ^{3}He polarization.

Safety is a priority in today's scientific environments. High-power lasers, heaters, and electro-magnets are possible hazards in SEOP. Various sensors will be installed in the ^{3}He NSF to monitor temperatures, electrical currents, voltages, cooling water for LDAs, and vibrations owing to earthquakes or accidental motion of the system. The sensor signals will be used for an interlock circuit to shut the system down automatically in case of an emergency. The ^{3}He NSF may be continuously operated for several weeks without maintenance on POLANO. Safety and reliability are crucial aspects in ensuring the success of the spectrometer.

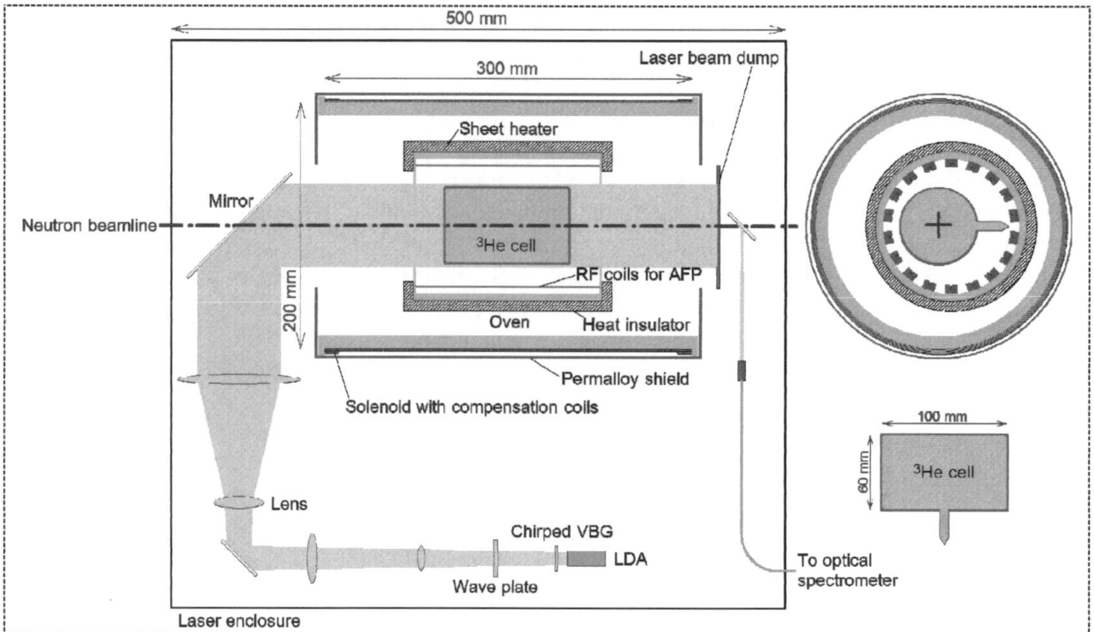

Figure 10. Cross-sectional view of the ^{3}He neutron spin filter on POLANO. The length is 500 mm along the beam direction, the width is 500 mm, and the height is 300 mm.

References

[1] T. Yokoo *et al.*, J. Phys.: Conf. Ser. **502**, 012046 (2014); K. Ohoyama *et al.*, J. Phys.: Conf. Ser. **502**, 012051 (2014); T. Yokoo *et al.*, J. Phys. Soc. Jpn. **82**, SA035 (2013); K. Ohoyama *et al.*, J. Phys. Soc. Jpn. **82**, SA036 (2013).
[2] T. G. Walker and W. Happer, Rev. Mod. Phys. **69**, 629 (1997); M. A. Bouchiat, T. R. Carver, and C. M. Varnum, Phys. Rev. Lett. **5**, 373 (1960).
[3] C. Y. Jiang *et al.*, Rev. Sci. Instrum. **85**, 075112 (2014); H. Kira *et al.*, Phys. Proc. **42**, 200 (2013); Sungman Lee *et al.*, J. Korean Phys. Soc. **62**, 419-423 (2013); E. Babcock, S. Mattauch, and A. Ioffe, Nucl. Instr. and Meth. A **625**, 43 (2011); S. Boag *et al.*, Physica B **404**, 2659–2662 (2009), and references therein.
[4] W. C. Chen *et al.*, J. Phys.: Conf. Ser. **528**, 012014 (2014); E. Babcock *et al.*, J. Phys.: Conf. Ser. **528**, 012018 (2014); C. Y. Jiang *et al.*, Rev. Sci. Instrum. **84**, 065108 (2013); S. Boag *et al.*, J. Phys.: Conf. Ser. **528**, 012019 (2014); Y. Sakaguchi *et al.*, J. Phys.: Conf. Ser. **294**, 012004 (2011); A. K. Petoukhov *et al.*, Physica B **385–386**, 1146 (2006); S. Lee *et al.*, "In-situ Compact ^{3}He Neutron Spin Polarizer Based on a Magneto-Static Cavity with Built-in NMR Coils," J. Phys.: Conf. Ser., this issue.
[5] N. R. Newbury, A. S. Barton, G. D. Cates, W. Happer, and H. Middleton, Phys. Rev. A **48**, 4411 (1993).
[6] H. Kira *et al.*, "Design of Cylindrical Glass Cells for A SEOP Based Polarized ^{3}He Neutron Spin Filter at J-PARC," J. Phys.: Conf. Ser., this issue.
[7] G. D. Cates, S. R. Schaefer, and W. Happer, Phys Rev. A **37**, 2877 (1988); L. D. Schearer and G. K. Walters, Phys. Rev. **139**, A1398 (1965).
[8] T. Ino *et al.*, J. Phys.: Conf. Ser. **340**, 012006 (2012).

[9] M. V. Romalis and G. D. Cates, Phys. Rev. A **58**, 3004 (1998).
[10] T. Ino *et al.*, Physica B **404**, 2667 (2009).
[11] E. Babcock, B. Chann, I. A. Nelson, and T. G. Walker, Appl. Opt. **44**, 3098 (2005); B. Chann *et al.*, J. Appl. Phys. **94**, 6908 (2003); I. A. Nelson, B. Chann, and T. G. Walker, Appl. Phys. Lett. **76**, 1356 (2000); B. Driehuys *et al.*, Appl. Phys. Lett. **69**, 1668 (1996).
[12] W. C. Chen, T. R. Gentile, Q. Ye, T. G. Walker, and E. Babcock, J. Appl. Phys. **116**, 014903 (2014); B. Chann *et al.*, Opt. Lett. **31**, 1253 (2006); C. Moser and G. Steckman, Photon. Spectra **36** (6), 82 (2005); B. L. Volodin et al., Opt. Lett. **29**, 1891 (2004).
[13] Y. Arimoto, et al., Physica B 406, 2439 (2011).

Measuring gas temperature during spin-exchange optical pumping process

E Normand[1, 2], C Y Jiang[1], D R Brown[1], L Robertson[1], L Crow[1], and X Tong[1, *]

[1] Instrument and Source Design Division, Oak Ridge National Laboratory, Oak Ridge, Tennessee 37831-6393, USA
[2] Tennessee Tech University, Cookeville, TN 38505, USA

Abstract. The gas temperature inside a Spin-Exchange Optical Pumping (SEOP) laser-pumping polarized ^{3}He cell has long been a mystery. Different experimental methods were employed to measure this temperature but all were based on either modelling or indirect measurement. To date there has not been any direct experimental measurement of this quantity. Here we present the first direct measurement using neutron transmission to accurately determine the number density of ^{3}He, the temperature is obtained using the ideal gas law. Our result showed a surprisingly high gas temperature of 380°C, compared to the 245°C of the ^{3}He cell wall temperature and 178°C of the optical pumping oven temperature. This experiment result may be used to further investigate the unsolved puzzle of the "X-factor" in the SEOP process which places an upper bound to the ^{3}He polarization that can be achieved. Additional spin relaxation mechanisms might exist due to the high gas temperature, which could explain the origin of the X-factor.

1. Introduction

Many neutron scattering facilities routinely use polarized ^{3}He neutron spin filters as neutron polarizers or analyzers [1-5]. Two main methods to produce polarized ^{3}He are metastability-exchange optical pumping (MEOP) and spin-exchange optical pumping (SEOP). Specifically in SEOP, alkali atoms, typically Rb and K in vapor form inside a glass cell (optical pumping cell) are first polarized through optical pumping. Then, via binary collisions between the alkali electrons and the ^{3}He atoms, the polarization is transferred to ^{3}He. However, optically-pumped alkali atoms can re-emit photons which can significantly reduce the polarization. To suppress this effect, a small amount of N_2 gas (50-100 torr) is added to the gas chamber to quench the excited alkali metal electrons non-radioactively back to the ground state by depositing the excitation energy to the rotational and vibrational states of the N_2 molecules [6, 7]. This energy can greatly increase the gas temperature inside the pumping cell which directly affects several key parameters in SEOP: the alkali metal vapor density, the alkali spin destruction rate and the alkali-^{3}He spin-exchange rate. In a conventional setup, the temperature is measured by placing a resistance temperature detector (RTD) or a thermocouple (TC) on the OUTSIDE wall of the glass cell; hence giving a reading that may not be the gas temperature inside the cell. Thus, a precise measurement of the temperature INSIDE the pumping cell is called for. Previous attempts include Raman spectroscopy [8] and NMR measurements of ^{3}He diffusion coefficient to determine the temperature inside the pumping cells [9]. These methods have shed light on temperature distribution and heat flow inside a pumping cell. However, none of these are direct measurements.

[*] tongx@ornl.gov

Assumptions and models need to be made to calculate the temperature. In this proceeding, we report a direct temperature measurement of a ^{3}He pumping cell under normal optical pumping conditions using a double-cell configuration and neutron transmission measurements. As far as we know, this is the first attempt to directly measure the temperature of the ^{3}He gas inside polarizing cells.

2. Experimental technique, Design and Construction

The goal of this experiment is to determine the temperature inside the ^{3}He cell being optically pumped. The idea is to use neutron transmission measurements to deduce the ^{3}He number density which in turn gives the corresponding temperature if the ^{3}He gas pressure is also known. The ideal gas law gives

$$P = nk_BT \tag{1}$$

where P is the pressure, n is the gas number density, k_B is the Boltzmann constant and T is the temperature. Once a ^{3}He cell is filled and sealed, the ^{3}He number density stays constant. For unpolarized neutrons passing through an unpolarized ^{3}He cell, the transmission is given by

$$T_n = T_e e^{-n\sigma(\lambda)L} \tag{2}$$

where T_n is the neutron transmission passing through ^{3}He, T_e is the neutron transmission through the container's glass wall, n is the ^{3}He number density, $\sigma(\lambda)$ is the neutron absorption cross section which is a function of the neutron wavelength λ and L is the inner length of the ^{3}He cell. From the neutron transmission, the ^{3}He number density can be deduced. However, for a single sealed cell there is no direct way to accurately measure the pressure when the cell is being pumped by lasers. Therefore, a double-cell setup was constructed as shown in Figure 1a. Both the top and bottom cells are made of GE-180 glass. The bottom cell (3" outer diameter) was placed in an oven that controls the alkali vapour pressure and optically pumped, i.e. the cell is heated while the laser light shines from both sides. The top cell (2" outer diameter) is outside the oven and does not see any laser light. The two cells are connected by a 0.5 inch diameter glass tube. It is 5 inches in length to make sure that the top cell is well away from the oven. Neutron transmission measurements can independently be done on the top or bottom cell. A thermocouple was inserted and epoxied into the glass tube near the opening of the top cell to give a direct measurement of the temperature in the top cell (Figure 1b). The thermocouple was not placed in the middle of the top cell in order to avoid interference with the neutron transmission measurements. Two additional thermocouples were taped on the outer wall of the bottom cell and the glass tube to read the wall temperature. Because the thermocouple on the glass tube was away from the pumping laser, it was a good indicator of the overall oven temperature. With the double-cell configuration, the ^{3}He number density can be determined for the top cell from the neutron transmission measurements. The temperature of the top cell can be measured with the inserted thermocouple. Then using Eq.1, the pressure of the top cell can be determined. To a good approximation, the pressure of the top cell should be equal to the pressure of the bottom cell at equilibrium. Therefore, once the number density of the bottom cell is determined by neutron transmission measurement, the temperature of the bottom cell can be calculated.

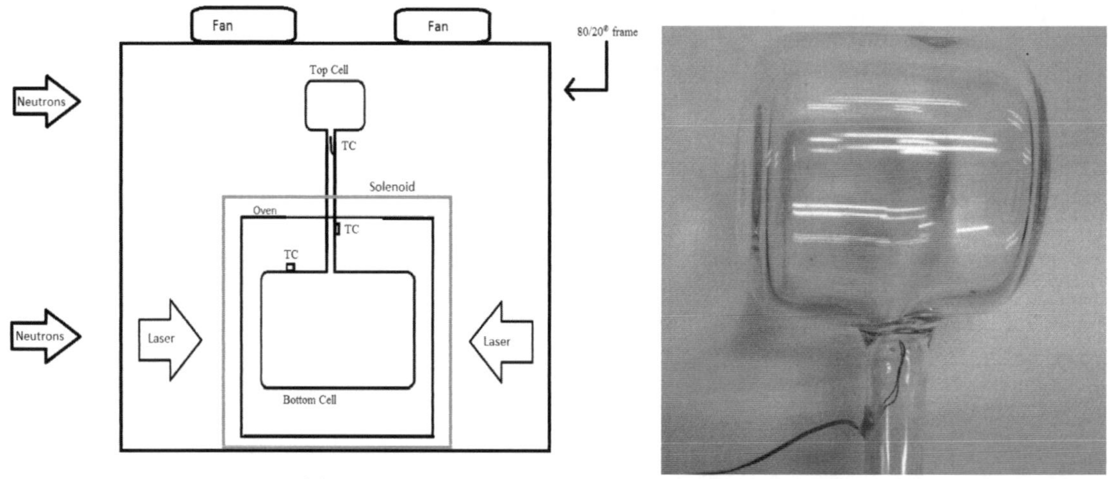

(a) **(b)**

Figure 1. **(a)** Schematics of the double cell system; **(b)** the thermocouple inserted through the glass stem into the top cell.

The oven that houses the bottom cell is made of G7 (glass silicone laminate) which has a temperature rating up to 220°C. On the front and back side of the oven two 2" holes were cut and covered with sapphire windows for neutrons and laser light to pass through. Instead of using compressed air to heat up the oven, electrical heating pads were chosen for simplicity and convenience [10]. The heating pads are PID controlled which helps stabilize the final temperature reading. Figure 3a shows the oven with the double cell and the heating pads in place. The two solenoids used were identical and placed over the oven so that the stem of the double cell would safely fit. Each solenoid has an 8.5" x 8.5" cross section and is 6" in length. Each one consists of 150 turns of 18 AWG solid copper wire. Current passes through the coils in series to produce a field at 15 Gauss.

(a) **(b)**

Figure 3. **(a)** The cell in the oven with heating pads in place; **(b)** the oven in the solenoid.

The system's base has a cut to fit optical breadboard that has ¼"-20 tapped holes in a 1" spacing pattern across its surface. This layout, as well as the use of a 200W fiber-coupled laser (0.35nm spectral linewidth), results in more simplified optics which simplifies the laser alignment (Figure 4a). After assembly, the dimension of the system is measured to be 61 cm in length along the axis of the neutron beam, 58 cm in width and 46 cm in height. The 80/20® outer frame used has enough space to house all of the optics and the solenoid, while still being compact enough to fit safely on the beamline table. The system was enclosed by laser safety panels with interlock switches on every side of the cube shape (Figure 4b). The interlock switches are set to shut down the laser system if any of the panels are removed.

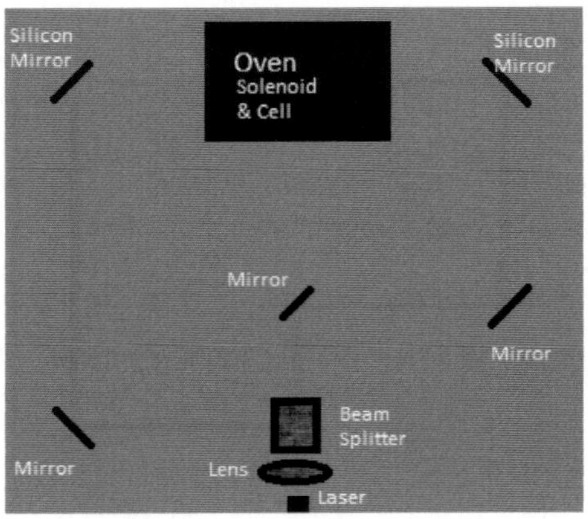

(a)

(b)

Figure 4. (a) Optical layout of the double cell system; **(b)** the enclosed system.

3. Tests and Results

The double cell was prepared and filled using the lab-based filling station [11] at the Oak Ridge National Laboratory. The cell was filled with 653 torr of ^3He and 47 torr of N_2. Rubidium was driven into the bottom cell only. The heating pads, optics, and laser were all tested in the lab before transferred to the polarized neutron development beamline HB-2D at High Flux Isotope Reactor (HFIR). At the beamline the system was mounted onto a hydraulic height-adjustable optical table so that transmission measurements could be performed for both the top and the bottom cell. This setup at the beamline is shown in Figure 5.

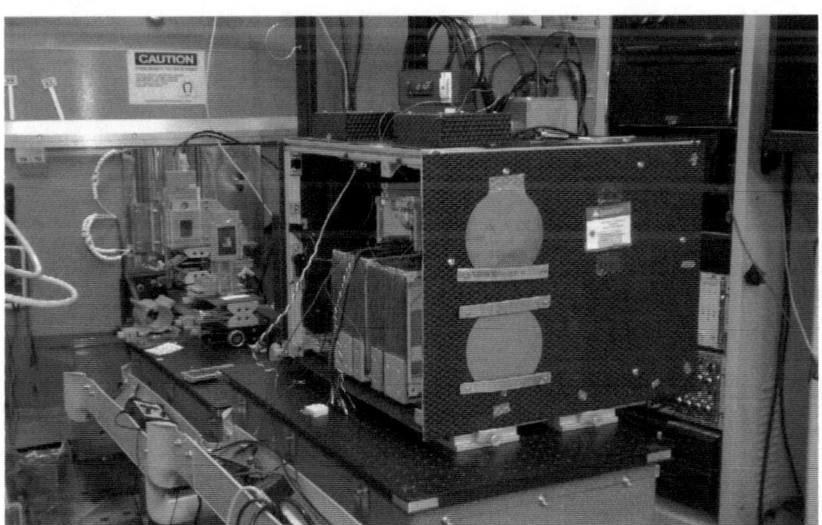

Figure 5. The system setup at beamline HB-2D at HFIR. The system was sitting on an optical table. One side panel was removed for illustration purpose.

In order to determine the temperature of the bottom cell, several neutron transmission measurements were performed: direct beam transmission without the cell, transmission with the cell at room temperature, transmission with the heating pads on and the laser off and transmission with both the heating pads and the laser on. These measurements were repeated for both the top and bottom cell.

The direct beam transmission was used to get a baseline reading. The neutron transmissions with the cell at room temperature were used to determine the inner length (IL) of the cell. The neutron transmission with the heating pads on and the laser off was used to determine if the calculated temperature and the measured temperature were in good agreement with each other. In this setting, the difference between the bottom cell wall temperature and the temperature inside the bottom cell should be negligible. Once it was determined that the calculated and measured temperatures agreed, neutron transmission measurements with the laser on full power were done to determine the temperature of the bottom cell. All of these measurements were performed when the temperatures in both the top and the bottom cell have stabilized. Table 1 shows a summary of the test results.

Table 1. The IL of both the top and bottom cell was calculated based on the room temperature neutron transmission runs. The calculated bottom cell temperature was very close to the measured cell wall temperature when the laser was off. With the laser set to full power (200W), the wall temperature and the calculated temperature of the bottom cell showed a large difference.

	Top Cell trans.	Bottom Cell trans.	IL Top Cell (cm)	IL Bottom Cell (cm)	T Top Cell (°C)	T Bottom Cell Wall (°C)	T Glass Stem (°C)	T Bottom Cell Calculated (°C)
Room Temperature	0.307	0.121	3.9	7.3	23.6	23.3	26.0	26.0
Heater on Laser off	0.250	0.130	3.9	7.3	31.5	109	102.5	105
Heater/Laser on	0.184	0.219	3.9	7.3	34.8	245.3	178	380

With the laser off, in which the spin-exchange process doesn't take place, the temperature read by the thermal couple on the bottom-cell wall should be close to the temperature of the gas temperature inside the cell. From table 1, the calculated and measured temperature in the bottom cell indeed showed good agreement which showed the validity of applying the ideal gas law to ^{3}He. In addition, as expected, the oven temperature taken by the temperature reading on the glass stem was also close to the measured and calculated bottom-cell temperature. With the laser on at full power, spin-exchange process took place and the quenching gas N_2 would heat up the internal gas substantially. The measurements revealed that the bottom cell temperature was elevated by about 135 °C compare to the bottom cell wall temperature, and by almost 200 °C compare to the oven temperature. The temperature change in the top cell was small compared to the change in the bottom cell.

4. Conclusion

We successfully demonstrated a direct ^{3}He temperature measurement using neutron transmission. A 135-200 °C temperature increase has been observed in our measurements. This confirms that the temperature inside the cell is higher than the oven temperature due to reemission heating from the quenching gas. To our knowledge, this is the first time that this temperature difference has been experimentally confirmed. How such an increase of temperature would affect spin-destruction and spin-relaxation of polarized ^{3}He is still unknown and requires further investigation. Since our purpose was only to observe the heating phenomenon, the ^{3}He was not highly polarized. Further experiments will be carried out for more precision measurements, including the polarized ^{3}He temperature measurement and the ^{3}He lifetime measurement at high temperatures. We hope such an effort will help solve the mystery of the "X-factor" [12].

Acknowledgements
We would like to thank our glassblowers Jason Craig and Ali Vandergrift for their wonderful work. This work was sponsored by the Scientific User Facilities Division, Office of Basic Energy Sciences, United States Department of Energy. Oak Ridge National Laboratory is managed by UT-Battelle, LLC, for the US Department of Energy (DOE) under contract no. DE-AC05- 00OR22725.

References:

[1] Chen WC, Gentile TR, Erwin R, Watson S, Ye Q, Krycka KL, et al. 2014 *J Phys Conf Ser* **528** 012014
[2] Andersen KH, Chung R, Guillard V, Humblot H, Jullien D, Lelievre-Berna E, et al. 2005 *Physica B* **356** 103-8
[3] Lelievre-Berna E 2007 *Physica B* **397** 162-7
[4] Pajerowski DM, Garlea VO, Knowles ES, Andrus MJ, Dumont MF, Calm YM, et al. 2012 *Phys Rev B* **86** 054431
[5] Ioffe A, Babcock E, Pipich V, Radulescu A 2011 *J Phys Conf Ser* **294** 012013
[6] Walker TG, Happer W 1997 *Rev Mod Phys* **69** 629-42
[7] Happer W 1972 *Rev Mod Phys* **44** 169-&
[8] Walter DK, Griffith WM, Happer W 2001 *Phys Rev Lett* **86** 3264-7
[9] Parnell SR, Deppe MH, Ajraoui S, Parra-Robles J, Boag S, Wild JM 2010 *J Appl Phys* **107**
[10] Tong X, Pierce J, Lee WT, Fleenor M, Chen WC, Jones GL, et al. 2010 *International Conference on Neutron Scattering 2009* **251**
[11] Jiang CY, Tong X, Brown DR, Lee WT, Ambaye H, Craig JW, et al. 2013 *Physcs Proc* **42** 191-9
[12] Babcock E, Chann B, Walker TG, Chen WC, Gentile TR. *Physics Review Letters*, 2006, **96**, 083003

Journal of Physics: Conference Series **711** (2016) 012013 doi:10.1088/1742-6596/711/1/012013

First result from the magic-PASTIS using large ^{3}He SEOP-polarized GE180 doughnut cell

Zahir SALHI[1], Earl BABCOCK[1],Ramil GAINOV[2,3], Klaus BUSSMANN[4],Hans KAEMMERLING[4], Patrick PISTEL[4], Margarita RUSSINA[2],Alexander IOFFE[1]

E-mail: z.salhi@fz-juelich.de

[1]Jülich Centre for Neutron Science; Forschungszentrum Jülich GmbH outstation at MLZ Lichtenbergstrasse 1 85747 Garching Germany.
[2]Helmholtz-Zentrum Berlin für Materialien und Energie GmbH,14109 Berlin, Hahn-Meitner Platz 1, Germany.
[3]Kazan Federal University, 420008 Kazan, Russia
[4]Forschungszentrum Jülich GmbH, ZEA-1Wilhelm-Johnen-Strßae 52425 J¨ulich Germany.

Abstract. We report on the first results of the newly proposed and prototyped PASTIS coil set, enabling for XYZ polarization analysis on the future thermal time-of flight spectrometers. Our setup uses a wide-angle banana shaped ^{3}He Neutron Spin Filter cell (NSF) to cover a large range of scattering solid angle. The design assures relative magnetic field gradients < 10^{-3} cm^{-1} and large solid angle areas not interrupted by either coils or supports. In the vertical direction nearly 40° are open and the blind spots in the horizontal scattering plane comprise only 3° in 180° due to the square X and Y compensation coils. We present the first results of the field mapping and relaxations time measurements using a large ^{3}He SEOP polarized GE180 doughnut cell.

1. Introduction

Polarization analysis, PA, of polarized neutrons is a powerful tool for separation of nuclear spin-incoherent background, analysis of complex magnetic structures and the study of magnetic excitations [1]. Several wide-angle spectrometers with PA exist or are under construction [2 -5]. The PA can be performed in a variety of ways depending on the instrument's parameters, however with performance limited by the analyzer height and, in some instances, integration over the height of the detectors or loss of resolution in the vertical direction due to the nature of a supermirror style analyser arrays. Installation of a new longer, height-position sensitive detector bank gives a unique opportunity to prototype and test. A polarized ^{3}He XYZ analysis system would be able to utilize the full height and position resolution of large area position sensitive detectors and enable analysis at higher neutron energies (energy transfer) than currently available[5 - 7].

For polarized ^{3}He-based systems, a sufficiently low magnetic field gradient must be maintained over the volume of the neutron spin filter cell in order to obtain ^{3}He polarization lifetimes of a practical and useful level. Further for any method to obtain information on the $Q_\perp$ and $Q_\parallel$ of the momentum transfer vector Q

over a large solid angle detector, the sample field must be switchable between three orthogonal X-Y-Z directions [9]. Due to practical constraints concerning the amount of polarized ^3He that can be provided, a so-called PASTIS (**P**olarization **A**naly**S**is on **T**hermal **I**nelastic **S**pectrometer) concept was developed [8], where the ^3He analyser cell shares the magnetic field of the sample. Since the cell occupies the same field volume as the sample, the direction of a large-volume highly uniformed ^3He holding field used to maintain long ^3He NSF lifetimes, must also be changeable between the X, Y and Z directions. Moreover, the ^3He polarization must rotate adiabatically when switching field directions, and the neutron polarization must be adiabatically preserved regardless of field direction. To achieve this, two clear options exist: a magnetized mu-metal geometry, similar to [8], or a resistive coil set similar to [6]. We use a geometry based on magnetized mu-metal. However in our proposed designs certain key differences exist, which build on the experiences from prior devices.

We present an initial design study with finite element magnetic field (FEM) calculations of possible XYZ field configurations suitable for polarized ^3He and adapted to the instrument geometry. Starting with the ideal geometry from the FEM model, the new setup was prototyped and tested after being optimized experimentally for current values and optimal positioning of the coils.

2. Influence of the magnetic field inhomogeneity on the polarized gas

The motivation of this work is the need for XYZ magnetic cavity, which provides a highly homogeneous magnetic field since the magnetic field gradients contribute strongly to the total relaxation of polarized ^3He gas [11-17]. In the presence of such gradients, a moving atom will experience different magnetic field strengths. If the variations of field are close to the transition frequency between Zeeman energy level for the diffusing atom, then spin flip can be inducted causing a loss of magnetization. T_{Grad} (gradient relaxation time) is the relaxation time due to the diffusion of ^3He through magnetic field gradient and given by [15]:

$$1/T_{1Grad}[\text{h}] = \frac{7000}{P[\text{bar}]}\left(\left(\frac{\Delta B_X}{B}\right)^2 + \left(\frac{\Delta B_y}{B}\right)^2\right) \quad (1)$$

Here p is the gas pressure in bar, ΔB_x and ΔB_y are the orthogonal gradients of the static holding field B. The influence of the magnetic field gradient on the total relaxation time for different pressures is shown in figure 1.

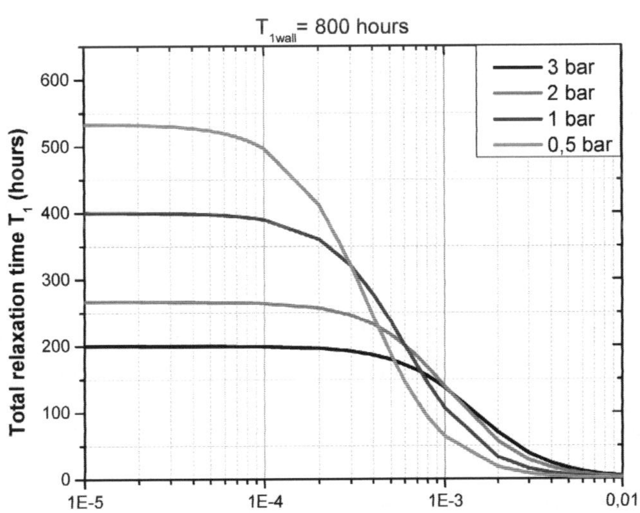

Figure1. Influence of the field gradient and gas pressure on the total relaxation time of polarized gas.

From Figure1, we can see clearly that the total relaxation time suffers strongly from the inhomogeneity of the magnetic field and is more sensitive to the field gradient at low pressure. For the 0.5bar pressure the observed saturation is about 8×10^{-5} cm^{-1}, whereas a significant decrease of the relaxation time at 3bar can be observed only for gradients larger than 10^{-3} cm^{-1}.Therefore, for the test of our setup we use a large cell with the pressure of 0.5bar in order to be more sensitive to the field gradient. For the final test with thermal neutrons, we will use a cell with an opacity of about 30 (10cm*3bar*1A) [25] to lower the influence of the gradient on the total relaxation time. Going to this higher pressure not only reduces the field homogeneity requirements, but also scales down the overall size of the cell for the thermal beam. This is important for two reasons: first, the total optical pumping efficiency is strongly related to the total volume of alkali-metal vapour that must be polarized, so that going to a smaller higher pressure cell, will in general reduce the amount of light required to polarized each ^{3}He atom, because the alkali-alkali relaxation mechanism tend to be dominate over alkali-^{3}He for our cell parameters. Second, the smaller cell will require a correspondingly smaller homogeneous magnetic field region, allowing this device to be more compact and suitable to the typical constraints on a neutron instrument.

3. Magnetic system design and prototype

A large step has been made in the design of wide-angle polarization analysis. The primarily PASTIS system developed at the ILL[8], uses a set of X, Y and Z coils to create a homogeneous magnetic field in three orthogonal directions in a compact format, which was suitable to known neutron instrumentation and had acceptable ^{3}He relaxation times. This device was further refined to reduce the dead angles [5]; however dead angles remain still the main limitation.

Our so-called Magic-PASTIS design [18] is based on a nearly open mu-metal geometry. The conceptual design was then adjusted and optimized using finite element calculations performed with commercially available software [19]. Some results of these calculations are shown in Figure 2. The present coil layout consists of a pair of compensated Helmholtz coils coupled to horizontal mu-metal sheets, which work as a kind of magnetic capacitor and generate a field in the Z direction. Two further pairs of rectangular coils (together with current sheets, realized by two orthogonal coils wound upon the mu-metal sheets, and magnetized iron-core solenoid rods) are used to generate a homogeneous field along X and Y directions.

We aim to produce an improved homogeneous magnetic field in a large region by introducing mu-metal sheets while reducing the blind areas due to the coil frames of the more conventional compensated square/circular coils of the resistive coil PASTIS systems [6,8]. Those systems, for example, have eight vertical dead areas (in the horizontal plane) of approximately 5° or more at ±18°; ±71°and ±109° whereas the design presented in this article has more than 40° opening in the vertical plane and only four small 3° dead angles, one every 90°, in the horizontal plane. These dead angles can be rotated arbitrarily with respect to the incident beam for either system. The centre of the configuration is the sample installed in a magnet or a cryostat, thus placing the ^{3}He cell itself significantly off-centre. A picture of the completed prototype is shown in Figure 3.

Another feature of our approach is that we have also fully modelled the neutron guide field configuration. Typically on PASTIS systems the in-plane X and Y directions provide relatively good neutron spin transport to the sample and through the ^{3}He NSF cell; however the vertical, i.e. transverse to the neutron beam, Z-direction proves to be more difficult as crossing through the side of a Helmholtz pair give a low field area for the neutron path. A guide field should then be added to this region, but this causes often to increase of ^{3}He relaxation rates because of the proximity to the cell. Therefore we have fully modelled the guide fields in conjunction with the magic-PASTIS system to insure that the polarization of both the neutrons and the ^{3}He will be properly preserved. A discussion of the guide field concept has been given in [10].

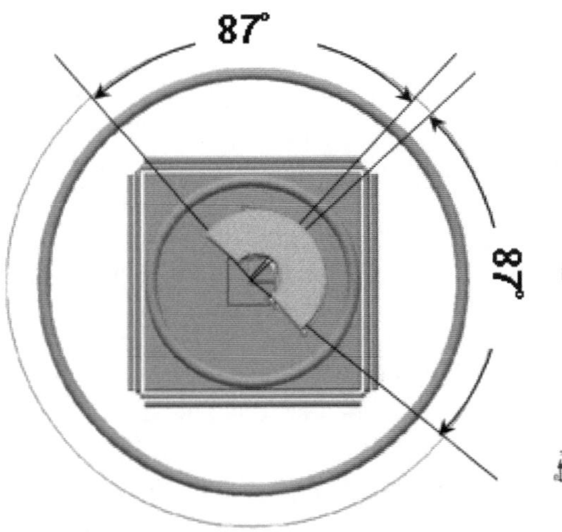

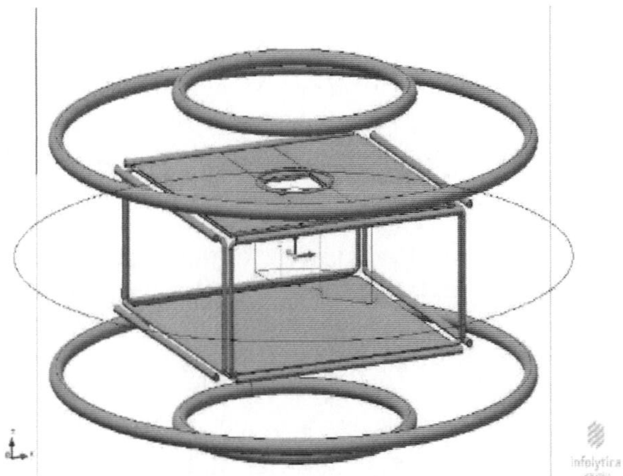

Figure 2. Coil layout for the PASTIS concept. Left: top view of the setup showing the cell position and the 3^0 dead angle. Right: a side view of the setup showing the opening for the cryostat and the cell position. Helmholtz coils (70cm), horizontal mu-metal sheets (35cm*35cm)

Figure 3.The magic-PASTIS prototype. Inside one can see a large doughnut-shaped ^{3}He cell (with NMR coils on the top of it) used for measurements of ^{3}He relaxation time.

4. Field mapping results

The magnetic field inside of Magic-PASTIS has been measured to optimize the homogeneity of the magnetic field in our set of coils. This has been done at different positions along the magnetic field axis using a Hall probe magnetometer Model 7030 manufactured by F.W. Bell [20], which has the sensitivity of 10nT. The computed currents obtained using the FEM calculations have been served as starting values. After that the field was measured at different ratios of electrical currents until the best field homogeneity was achieved.

Figure 4a shows FEM calculations of the magnetic field in the x, y direction. It is obvious that the inner part is highly homogeneous, including also the position of the ^{3}He cell. As seen from Fig. 2, the coil setup fulfils this condition for a large area, allowing the polarization analysis in a wide angular range. Figure 4b shows an example of the field map along the x, y direction achieved for the optimal current values. The measured magnitudes of magnetic fields are in good agreement with calculations. The values optimised here are used later for the measurement of the relaxation time (see section 5).

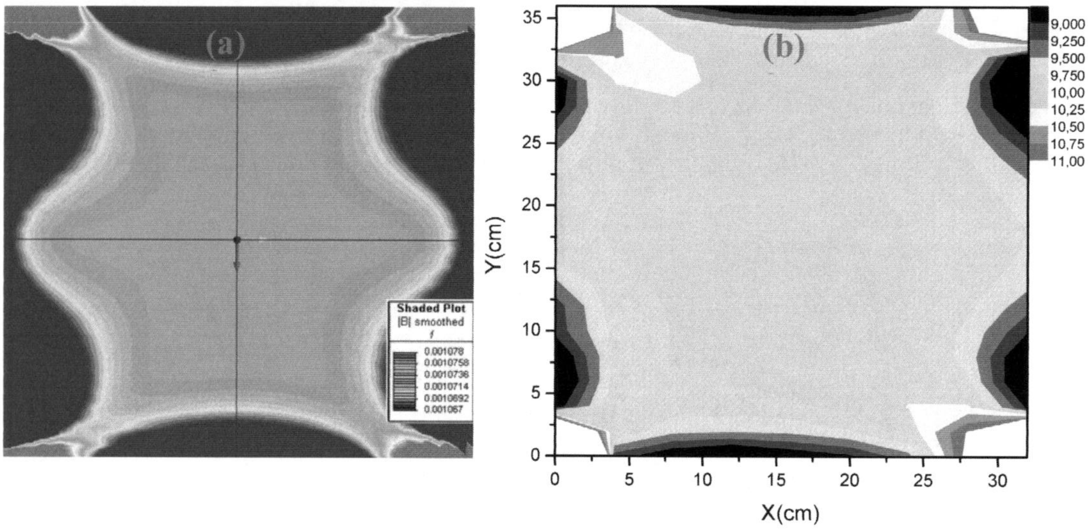

Figure 4.(a) FEM calculation for the X, Y directions,(b) magnetic field mapping along X, Y directions measured using Hall probe magnetometer.

5. Procedure for relaxation time measurement and results

Figure 5a shows the picture of a very large, D =22 cm, doughnut-shaped cell called Homer. The cell was prepared with Rb/K mixture and 0.5 bar of ^{3}He [21,22]. This cell is made purely only for testing of ^{3}He relaxation times in the new magic-PASTIS coil system shown earlier in Figure.1: the low ^{3}He pressure makes the cell very sensitive to magnetic field gradient relaxation and, as result, is a good probe to optimize the performance of the magic PASTIS magnetic cavity.
Aided by the low ^{3}He pressure, good glass and cell preparation techniques, Homer demonstrates 1050 hours of total relaxation time in what we presume to be an "ideal" magnetic holding field (Figure 5b). Regardless of the true lifetime of Homer in an actually "ideal" holding field and considering corrections for all losses from the NMR FID measurements of the lifetime, such a measured lifetime insures that with our NMR detection system, the measured lifetimes of Homer inside the magic-PASTIS system will be entirely dominated by the magnetic field gradients with only small corrections that we can neglect for the purpose of this work to optimise the field configuration.

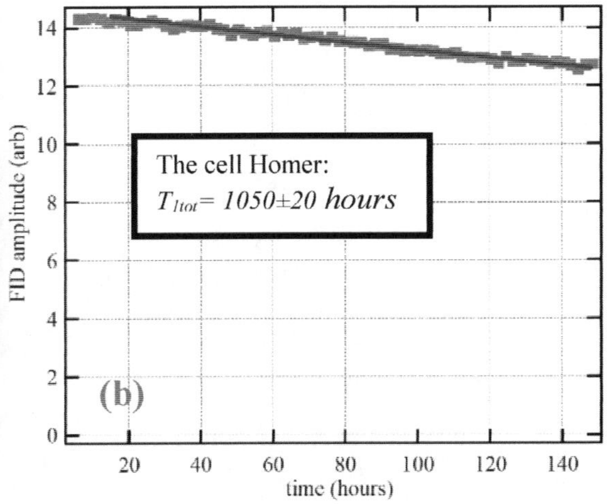

Figure 5.(a)First in the world doughnut shaped cell called Homer. The cell is filled with 0.5 bar of ^3He to make it sensitive to magnetic field gradients. (b)FID NMR exponential fit allowing for the determination of the relaxation timeT_1 = (1050 ± 20) hours;

The build-up and decay of the relative polarization was monitored using the NMR free induction decay (FID) [11-17]. The FID of the longitudinal magnetization was periodically monitored, typically every hour: ^3He polarization is proportional to the amplitude of the FID signal. The exponential decay fit of the amplitude-time dependence gives a value of T_1; for a long T_1 a reliable measurement typically takes several days.

The ^3He doughnut-shaped cell was polarized using the SEOP technique [23]. For this purpose the cell is placed in a long solenoid, providing the magnetic field of about 10G at the centre and the field homogeneity better than 1mG/cm along the axis at the centre of the solenoid coil. The cell is heated to achieve a suitable alkali metal density and illuminated by a circularly polarized light from a high power laser spectrally narrowed by the chirped volume Bragg grating (VBG). In order to cover the cell with polarized light the cell was illuminated from both sides: each laser covers the half of the cell. A large oven was built specially to host such large cell.

After reaching a reasonable polarization, the cell was transported in the permanently magnetised box with long gradient relaxation time [24], and, finally, placed in the centre of the Magic-PASTIS system as shown in Figure 3.

The results of the NMR FID measurements and the exponential fits allowing for the determination of the relaxation time T_1 are presented in Figures 6 and 7. Figure 6 shows ^3He lifetime measurements with Homer before and after optimization of correction coil currents for X direction. We point out here that T_1=70 hours obtained at 0.5bar using the testing cell corresponds to T_{1grad}=450 hours of magnetic lifetime and total T_{1tot}=165 hours for the case of a working pressure of 3bar. These results have been well reproduced by several measurements.

The same procedure has been also repeated for other two directions: Y and Z. The results are summarized in Figure 7. The obtained values of T_1 for X and Z directions are about 70h (450h at 3bar). Unexpectedly, the T_1 obtained for Y direction was only 35h in contrast to T_1=70h obtained for X direction. Regarding the symmetry of the x-y coils geometry, this can be explained by the fact that optimal currents values of the compensations coils for X direction are not appropriate for those of Y direction because of a small mechanical asymmetry or winding inaccuracy. More efforts could be done to obtain the appropriate optimal current ratios using field mapping and optimization. However, even the achieved relaxation time of T_1=35h at 0.5bar corresponds to T_{1grad}=220h and T_{1tot}=120h at 3bar, which is satisfactory for our applications. This value is still better than our aim to achieve the total relaxation time of 100h because of an important reason: to refill the cell by polarized ^{3}He gas not often than one time per day.

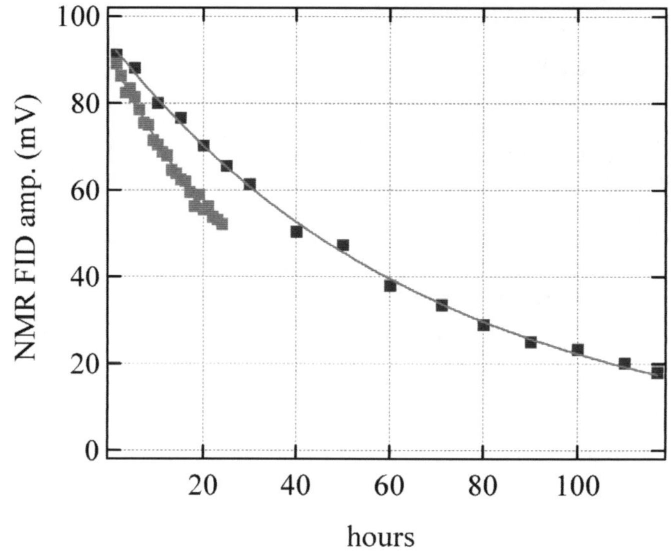

Figure 6. Results of ^{3}He lifetime measurements with Homer before (Red) and after (Blue) the optimization of correction coil currents for X direction. T_1= 70 hours obtained at 0.5bar corresponds to T_{1grad}=450 hours of magnetic lifetime and total T_{1tot}=165 hours at 3bar.

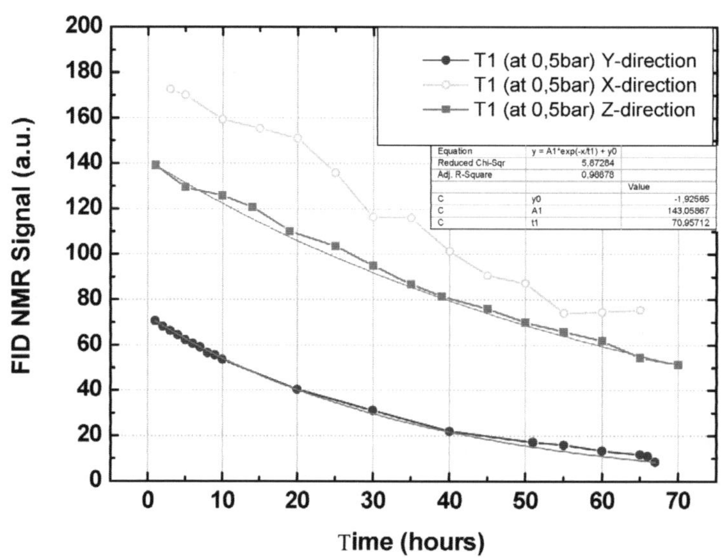

Figure 7. Results of ^{3}He lifetime measurements after the optimization of the correction coil currents for all three directions.

It is necessary that the field gradient is kept below 10^{-3}cm^{-1}over the volume of a large neutron spin filter cell in order to make the depolarization of the polarized ^{3}He sufficiently low. This level would keep the cells in the instrument during a long time, which is enough to make experiments feasible by reducing the time spent refilling, calibrating and transporting polarized cells/gas to the beamline to once per day or less. For our setup the gradient relaxation time was measured to be 450 h, corresponding to volume averaged field gradient of 8×10^{-4}cm^{-1}, which will allow us to work with the polarized cell without refilling for more than 24h with a good performance.

4. Conclusions

We have designed, prototyped, and tested the magic-PASTIS XYZ magnetic cavity. The design assures relative field gradients $<10^{-3}cm^{-1}$ and large solid angle areas, which are not interrupted by either coils or supports. In vertical direction nearly $\pm 20^{\circ}$ are open and the blind spots in the horizontal scattering plane comprise only 3^0 due to the square compensations coils for X and Y directions. The magic-PASTIS magnetic cavity prototype has been tested for ^3He lifetime, and a 70 hour lifetime at 0.5bar was achieved which corresponds to the 450 hour magnetic lifetime or the 165 hour total lifetime for a 3bar cell when including the ^3He dipole-dipole self-relaxation to the magnetic field gradient relaxation.

Work is steadily progressing towards the realization of optimized GE180 banana cells, the PASTIS control system and the guide fields for polarized neutrons. Once the critical mass of components is available we will proceed with neutron sample measurements, data collection and analysis techniques.

6.1. Acknowledgments

We acknowledge support from the BMBF through the collaborative project 05E10CJ1 of the ESS design update work package K3 and the project 05K12CB1 supported the cooperation between JCNS and HZB. This research project has been also supported by the European Commission under the 7th Framework Programme through the 'Research Infrastructures' action of the 'Capacities' Programme, NMI3-II Grant number 283883. The authors want to thank A. Schwaitzer (from the JCNS glass workshop) for his help and advices concerning cell fabrication.

References

[1] Moon R M, Riste T, Koehler W C., *1969 Phys. Rev. 181 920.*
[2] Stewart J R, Deen P P, Andersen K H *et al., 2009 J. of Appl. Crystal. 42 69-84*
[3] Schweika W, Boni P., *2001 Physica B 297 155*
[4] Voigt J, Babcock E, Brueckel Th., *J. of Phys. Conf. Series 211 012032*
[5] Beecham C J, Boag S, Frost C D *et al., 2011 Physica B- Condensed Matter* 406 2429-32
[6] Andersen K H, Jullien D, Petoukhov A K, Mouveau P, Bordenave F, Thomas F, Babcock E., *2008 Physica B-condensed matter 404 2652-4*
[7] Babcock E, and Ioffe A., *2011 Jornal of Phys. Conf. Series 294 01 2005*
[8] Stewart J R, Andersen K H, Babcock E, *et al. 2006 Physica B-condensed matter 385-86 11425*
[9] Scharpf O, and Capellmann H., *1993 Physica Status Solidi A - App. Res. 135 359-79*
[10] Voigt J, Soltner H, Babcock E, Aldus R J, Salhi Z, Gainov R R, Brückel T., *QENS/WINS2014, May 2014 Autrans France*
[11] Newbury N R, Barton A S, Cates G D, Happer W, Middleton H., *Phys. Rev. A, 48(6), 1993. doi:10.1103*
[12] Cates G D, Schaefer S R, Happer W., *Phys. Rev. A, 37, 1988.*
[13] Cates G. D, White D J , Chien T, Schaefer S R, Happer W., *Phys. Rev. A, 38, 1988.*
[14] Jacob R, Teter J, Saam B, Chen W, Gentile T., *Physical Review A, vol. 69 2004.*
[15] Schmiedeskamp J, Heil W, Otten E, Kremer R, Simon A, Zimmer J., *The European Physical Journal D, 2006*
[16] Jacob R E, Morgan S, Saam B, Leawoods J C, *Physical Review Letters, vol. 87, p. 143004, October 2001.*
[17] Babcock E D., *PhD Thesis, University of Winsconsin-Madison (2005)*
[18] Babcock E, Salhi Z, Pistel P, Simeoni G, Ioffe A., *Journal of Physics: Conference Series 528(2014) 01201*
[19] MagNet 2D/3D Electromagnetic Field Simulation Software, *Infolytica Co. Montral, Qubec H2X 4B3 Canada, http://www.infolytica.com/en/products/magnet/*
[20] http://fwbell.com/products/gauss-meters.aspx
[21] Salhi Z, Babcock E, Pistel P, Ioffe A., *Journal of Physics Conference Series 07/2014; 528(1):012015.*

[22] Chen W C, Gentile T R, Fu C B, Watson S, JonesG L, McIver J Wand RichD R, *Journal of Physics: Conference Series 294(2011)12003*
[23] Walker T, Happer W., Rev Mod Phys, vol. 69, no. 2, pp. 629–42, 1997.
[24] Petoukhov A K, Guillard V, Andersen K H, Bourgeat-Lami E, Chung R, Humblot H, Jullien D, Lelievre-Berna E, Soldner T, Tasset F, Thomas M., *Nuclear Instruments and Methods in Physics Research A 560 (2006) 480–484*
[25] Petoukhov A K, Andersen K H., et al *Physica B 2006 385-386 1146-1148*

In-situ compact ^{3}He neutron spin polarizer based on a magneto-static cavity with built-in NMR coils

S Lee[1], M K Moon[2], J Kim[2], S J Cho[2], J H Lee[3], C-H Lee[3], S W Lee[4], T Ino[5]

[1]Quantum Optics Division, Korea Atomic Energy Research Institute, Daejeon 305-353, Korea
[2]Neutron Instrument Development and Maintenance Division, Korea Atomic Energy Research Institute, Daejeon 305-353, Korea
[3]Neutron Science Division, Korea Atomic Energy Research Institute, Daejeon 305-353, Korea
[4]School of Mechanical Engineering, Pusan National University, Busan 609-735, Korea
[5]Neutron Science Laboratory, KEK, Tsukuba, Ibaraki 305-0801, Japan

E-mail: smlee3@kaeri.re.kr

Abstract. A polarized ^{3}He neutron polarizer for in-situ neutron beam line operation was developed based on a compact magneto-static cavity with a dimension of 280×270×300 mm^3 and a fiber-coupled VBG (Volume Bragg Grating) diode laser with a narrow spectral bandwidth of 25 GHz. Built-in NMR coils of the neutron spin polarizer designed for NMR signal measurements were described in detail and their performances were tested for monitoring the progress of in-situ ^{3}He polarization.

1. Introduction

Unlike neutron guides or neutron benders, which change the neutron beam alignment and/or neutron spatial beam uniformity owing to their characteristic multiple reflections, polarized ^{3}He neutron polarizers are free from such beam misalignment and beam interference [1]. In addition, a ^{3}He polarizer has a wide acceptance angle and creates a smaller amount of background gamma radiation owing to its low neutron activation [2]. Such merits enable the ^{3}He polarizer to be widely used for neutron experiments in fundamental particle physics or material researches involving neutron radiography, neutron diffraction, neutron reflectometry, neutron interferometry, and small-angle neutron scattering [3,4].

At KAERI, two programs were recently funded to develop a ^{3}He polarizer and a polarizing supermirror for a deployment of polarized neutrons in the neutron beam lines of the HANARO (High-flux Advanced Neutron Application ReactOr). A neutron guide test station G-TS will be utilized to test both types of polarizers. Currently, one of the top priorities is to apply the developed neutron polarizers for a neutron reflectometer (REF-V). In addition, the ^{3}He polarizer will be deployed for neutron imaging applications. For such applications, a polarizer and an analyzer will be needed and we are developing ^{3}He polarizers based on a compact solenoid coil or a compact magneto-static cavity.

In this paper, we report a ^{3}He-cell fabrication station and a magneto-static cavity based ^{3}He polarizer. In particular, the design of an in-situ polarizer including a NMR signal measurement technique and an optical pumping method is described in detail.

2. ^{3}He-cell Fabrication

A rectangular shape quartz cell was made for uses in magneto-static cavity based ^{3}He polarizer and a wide angle cell for a multi-axis spectrometer has been planned. In addition, for applications with a solenoid based ^{3}He polarizer in a neutron reflectometer, neutron imaging, SANS, TAS, etc., cylindrical high purity Ge-180 glass cells were fabricated with various sizes and gas pressures.

2.1. A high purity ^{3}He-cell

A high purity cylindrical Ge-180 glass cell with a length of 50 mm and a diameter greater than 50 mm was fabricated, as shown in figure 1(a) and 1(b), for uses in a solenoid-based ^{3}He neutron spin polarizer. The GE-180 glass cells were prepared at a glass shop in the nuclear chemistry division of KAERI, and then filled with ^{3}He gas by using a ^{3}He-gas filling station constructed in the quantum optics division of KAERI.

Figures 1(c) and 1(d) show a wide-angle quartz cell and a rectangular quartz cell made from a local glass shop. The wide-angle cell has a cross-section of 50×50 mm^2 and an angle of 120°. The size of the rectangular cell was 50×50×50 mm^3. After distilling rubidium metal vapor into the cell, the cell was filled with ^{3}He gas and nitrogen gas with a pressure ratio of 9:1. The cell was sealed off at a gas pressure of 0.9 atm, which can be obtained by immersing the cells in liquid nitrogen.

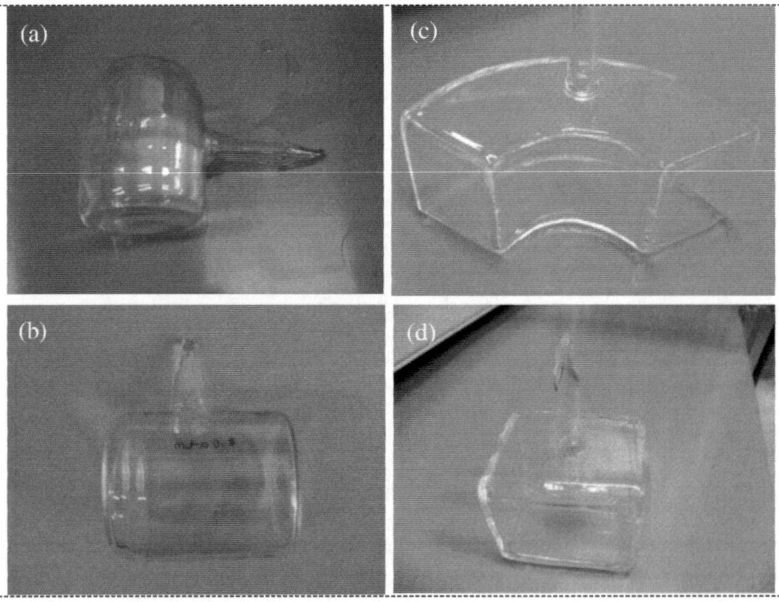

Figure 1. (a) Cylindrical ^{3}He gas cell made of Ge-180 glass (gas pressure: 1.5 atm), (b) cylindrical ^{3}He gas cell made of Ge-180 glass (gas pressure: 1.0 atm), (c) wide angle quartz cell (gas pressure: 1.2 atm), and (d) rectangular quartz cell (gas pressure: 1.0 atm).

2.2. Gas filling station

Figure 2 shows the constructed gas filling station system. The system was equipped with a high vacuum system including dry pump, turbo pump and ion pump, a high temperature cell-baking oven, a gas purification and inlet system, a residual gas analyzer, a manometer, and a liquid-nitrogen-cooling cell seal-off system, and band heaters for gas-pipe heating.

To obtain a high vacuum, all vacuum gas pipes made of stainless steel were heated using band heaters at a temperature of 100℃, and a glass cell and a glass manifold near the cell were baked using a high temperature oven operating at a temperature of 400℃. The remaining portion of the glass manifold made for rubidium metal vapor inlet into the cell was flame-baked with a hand torch until no vacuum pressure was changed during the flame heating. A small ion pump (VIPC1000, VMT) was used in combination with a dry and a turbo vacuum pump to obtain the desired high vacuum. With the vacuum system and flame heating, a high vacuum of 5×10^{-9} Torr was obtained within several days of operation.

Under high vacuum, a rubidium ample was broken by a hammer. The rubidium was vaporized by a flame torch, and then chased into the cell by using a torch and a heat gun. After chasing a small portion of rubidium into the cell, the cell was filled with nitrogen gas and then vacuumed for cleaning of the gas filling tubes. The ^{3}He gas had a purity of 99.995% (9.1 bar); and the N_2 gas, 99.9999%, and these gases were purified by gas purifiers before filling. The filling gas pressures of the cell were controlled using a needle valve and a baratron manometer (mks, 629D) which can measure the gas pressure from 20 mTorr to 5000 Torr. After obtaining the desired mixed gas pressures of the cell, the cell was immersed in a liquid nitrogen bath and cooled to a desired sealed-off gas pressure of 0.9 atm. During the sealed-off process of the cell, the gas pressure was maintained close to the sealed-off pressure by adjusting how deep the cell was immersed in liquid nitrogen.

To easily seal off the cell from the glass manifold, a GE-180 cell was fused with a GE-180 glass tube, which has a thin bottle neck. In addition, the cell was mounted in the liquid nitrogen bath on a lab jack. The height of the lab jack was adjusted according to the melting degree of the bottle neck. Then, the melted bottle neck was closed off due to the pressure difference between the cell and the atmosphere and the seal-off of the cell is completed. Using the seal-off procedure above, various high pressure cells can easily be manufactured.

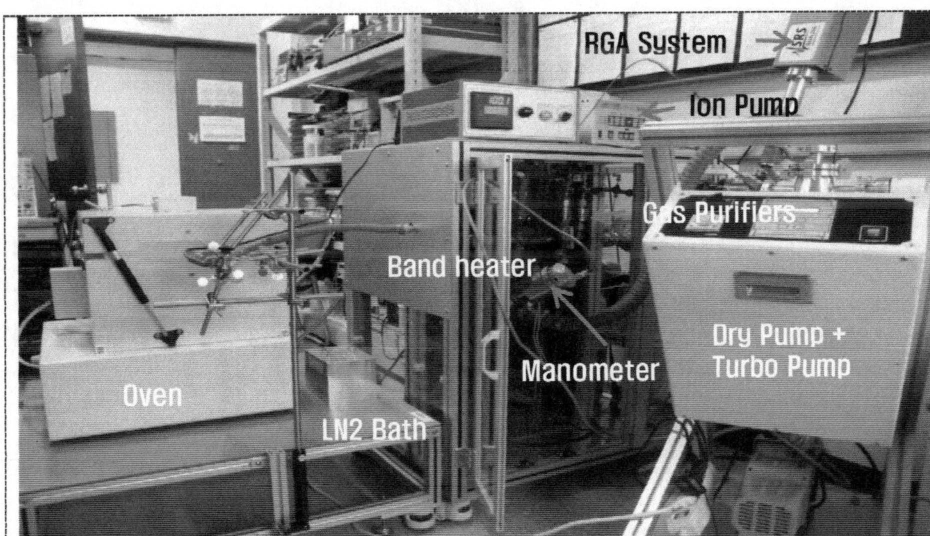

Figure 2. A ^{3}He-cell fabrication station constructed for sealing a high purity ^{3}He cell.

Journal of Physics: Conference Series **711** (2016) 012014 doi:10.1088/1742-6596/711/1/012014

3. Novel Magneto-static Cavity Based In-situ ^{3}He Polarizer

A unique in-situ ^{3}He polarizer based on a novel magneto-static cavity was developed. The magnetic field profile of the magneto-static cavity was analyzed using a commercial magnetic field calculation code, Vizimag 3.19. In addition, a NMR system was designed and installed inside the cavity to measure the polarization of the ^{3}He cell in real time.

3.1. Magneto-static Cavity

A ^{3}He polarizer takes at least 1 day to be polarizer through the Spin Exchange Optical Pumping process. It is often cumbersome to then transport the polarizer to neutron beam lines and the polarization ratio decays with time. Therefore, an in-situ neutron polarizer is preferred to utilize the polarizer more efficiently.

For use as an in-situ neutron polarizer, a magneto-static cavity was designed based on the analysis of magnetic field profiles depending on the magnet's size and strength, cavity's size and structure, and magnetic material properties. Figure 3 shows the magneto-static cavity with a small Helmholtz coil inside the cavity. The cavity consists of symmetric magnetic side-structures and symmetric top and bottom magnetic plates. The side-structure consists of a mild steel plate with a thickness of 3 mm and two arrays of a permanent magnet (N35) as shown in figure 3(right). The mild steel was magnetized by the strong permanent magnets. The magnetized mild steel was held by an inside aluminum frame, which was also maintained by top and bottom aluminum plates. On the top and bottom aluminum plates several permalloy magnetic sheets with high permeability are attached and connected with the magnetized mild steel of the side-structure. The top and bottom permalloy sheets are then magnetized and a uniform magnetic field is formed along the top-to-bottom direction at the central part of the cavity. The other mild steel plates are placed at the outside of side-structure with a gap from the magnetized mild-steel plates. By adjusting the gap distance, the inside magnetic field distribution and the strength of the cavity can be varied. The magnetic field strength can also be changed by the Helmholtz coil located in the vertical direction.

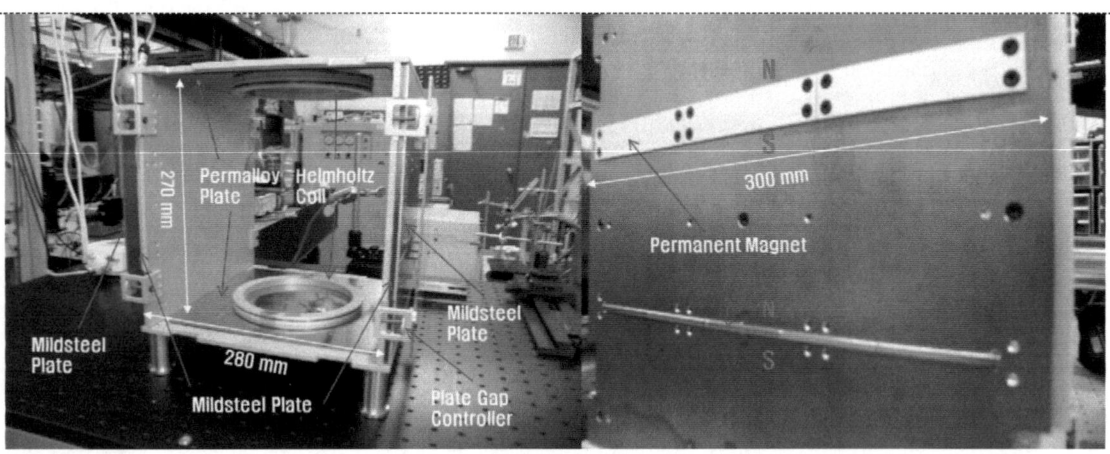

Figure 3. A magneto-static cavity: front view (left) and side view (right).

The magnetic field distribution of the cavity was measured as a function of the current of the Helmholtz coil using a magnetic field strength meter (FH 54, Magnet-Physik). Figure 4 shows the resulting magnetic field strength measured along the horizontal, vertical, and axial directions. A static magnetic field of 14 Gauss generated from the permanent magnet was increased to approximately 17 Gauss for an applied current of 1.8 A. These magnetic field distributions correspond to the results measured before a field adjustment using the outside mild plate gap controller, and a slight asymmetry in magnetic field profile can be compensated.

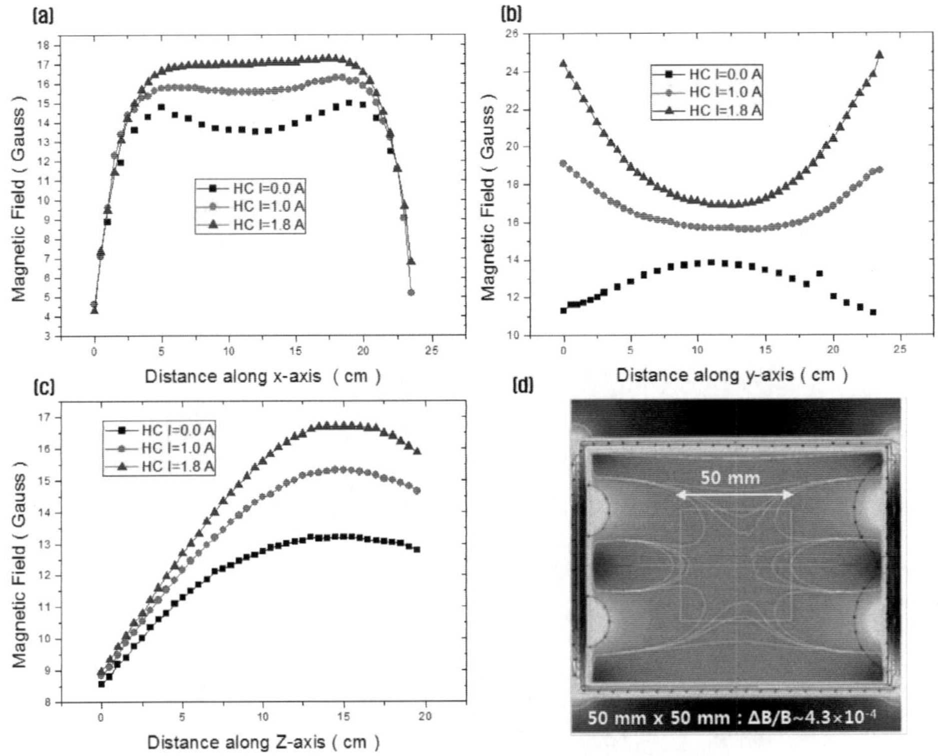

Figure 4. Magnetic field distribution of the cavity, which were measured along (a) the horizontal (x-axis), (b) vertical (y-axis), and (c) axial (z-axis) directions. The calculated magnetic field profile of the cavity is shown in (d). Zero positions in (a) and (b) are 10 mm apart from the side mild steel plate and bottom permalloy plates denoted in the figure 3, respectively, and that in (c) corresponds to the front end of the cavity.

In the figure 4(a), 4(b), and 4(c), the measured magnetic fields are the summed magnetics originated from the cavity and the Helmholtz coil. These magnetic fields are only used during a sweeping period for NMR signal measurements. To maintain the He-3 polarization, only the uniform magnetic field of the cavity can be used. The uniformity of the cavity's magnetic field, $\Delta B/B$ is calculated as $\sim 4.3\times10^{-4}$ within a cell size of 5 cm×5 cm. The uniformity can be improved by using a larger size cavity.

3.2. In-situ neutron spin filter based on magneto-static cavity

An in-situ ^{3}He neutron spin filter based on the novel magneto-static cavity was developed for use as an in-situ neutron polarizer or a neutron analyzer in the neutron beam lines of a HANARO research reactor. The in-situ neutron spin filter consists of the magneto-static cavity, a laser beam transport system, a ^{3}He cell, a nuclear magnetic resonance (NMR) signal measurement system, and a hot air supply system.

The magneto-static cavity had a width of 280 mm, a height of 270 mm, and a length of 300 mm. Two sets of permanent magnet are installed on each side. The ^{3}He cell is placed at the center of the cavity where a uniform magnetic field is formed. The ^{3}He cell was surrounded by a cosine theta NMR coil bobbin. The rubidium in the ^{3}He cell was vaporized by hot air at a temperature of approximately 150℃, which is supplied into the cosine theta coil by an external hot air supply system.

Journal of Physics: Conference Series **711** (2016) 012014 doi:10.1088/1742-6596/711/1/012014

A VBG (Volume Bragg Grating) diode laser with a narrow spectral linewidth of 18 GHz was used for pumping vaporized rubidium D_1 absorption lines at a wavelength of 794.8 nm. The precise wavelength tuning ability was obtained by a temperature control of the VBG external cavity. As shown in figure 5, the fiber delivered laser beam was collimated to have a large proper diameter and circularly polarized by a combination of a Galilean telescope, a polarizing beam splitter (PBS), and a $\lambda/4$-plate. The enlarged laser beam is reflected by an aluminum mirror (M_1) and then absorbed uniformly by the vaporized rubidium. The pumping laser beam was directed along a top-to-bottom direction of the cavity. Thus, the laser propagation direction was parallel to the direction of the magnetic field of the cavity and a selective σ^+ optical pumping of rubidium became possible. The transmitted laser beam can also be reabsorbed after being reflected back to the cell by the other aluminum mirror (M_2).

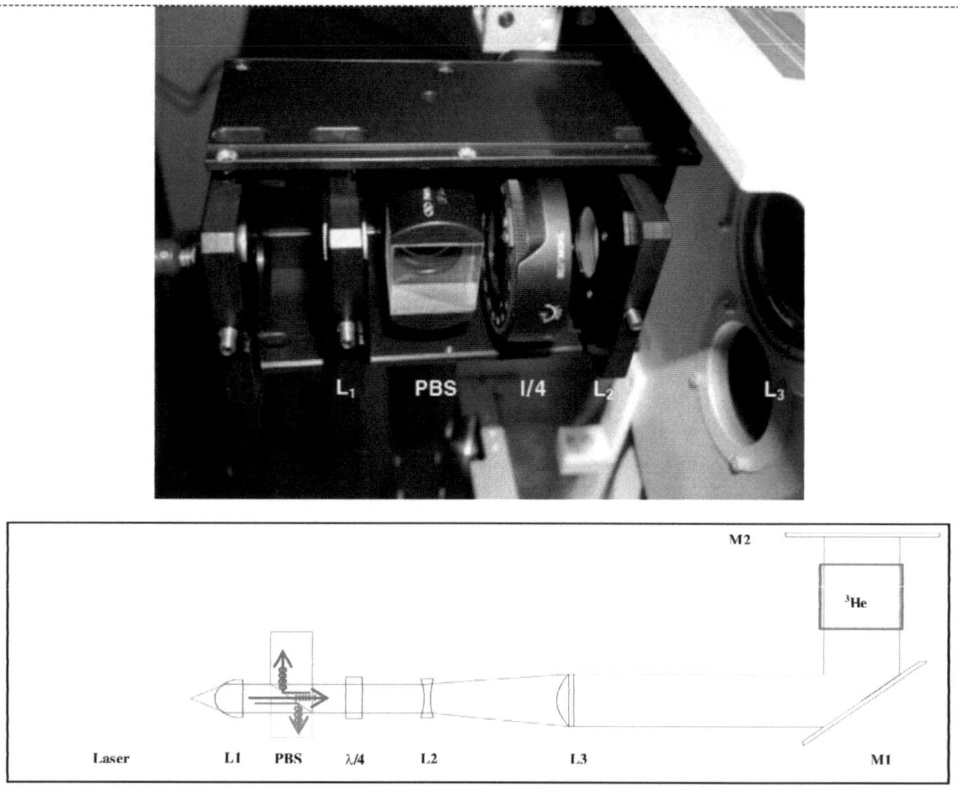

Figure 5. A compact optical pumping system (top) and its schematic diagram (bottom): L_1, L_2, L_3 (lens), PBS (polarizing beam splitter), $\lambda/4$ (quarter-wave plate)

The NMR coils were developed to measure the 3He polarization and were installed within the magneto-static cavity. As shown in figure 3, the Helmholtz coil was installed to have the same vertical magnetic field with that of the magneto-static cavity. The summed magnetic fields were used as an external holding magnetic field for the polarized ^{3}He gas. A triangular shape current was applied to the Helmholtz coil to sweep the Larmor precession frequency, which is proportional to the magnitude of the holding magnetic field. A cosine theta coil was used as a RF coil for NMR signal measurement [5]. As shown in figure 6, the magnetic field **B** of the RF coil was formed from left to right direction. The resonance frequency of the coil was tuned to a Larmor frequency (f=50 kHz) of ^{3}He gas for the given holding magnetic field.

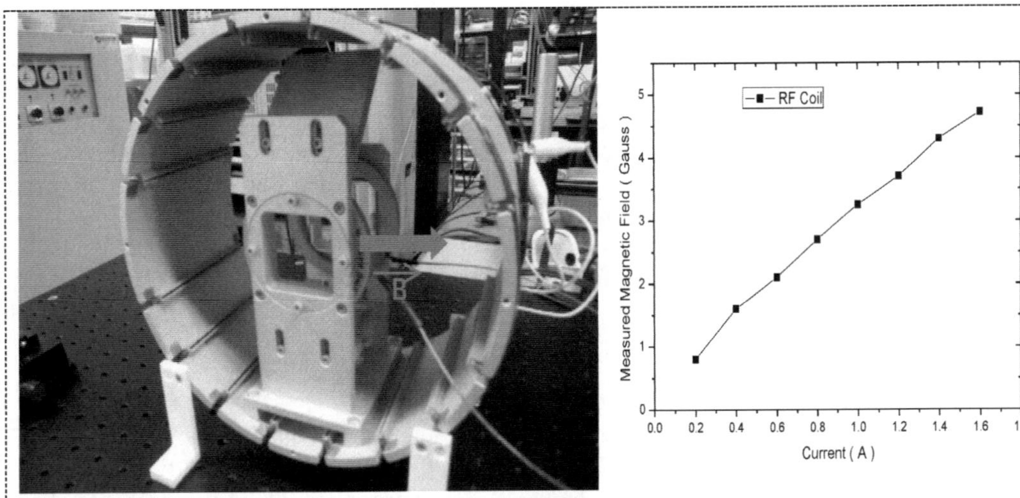

Figure 6. A cosθ coil used for the NMR signal measurement (left), and the magnetic field **B** measured at the varied dc current of the cosθ coil (right).

A ring-type pick-up coil has a magnetic field in the horizontal direction perpendicular to that of the cosine theta coil and was employed for NMR signal measurement. The resonance frequency of the pick-up coil was also adjusted to the Larmor frequency of the ^{3}He gas for the given holding magnetic field.

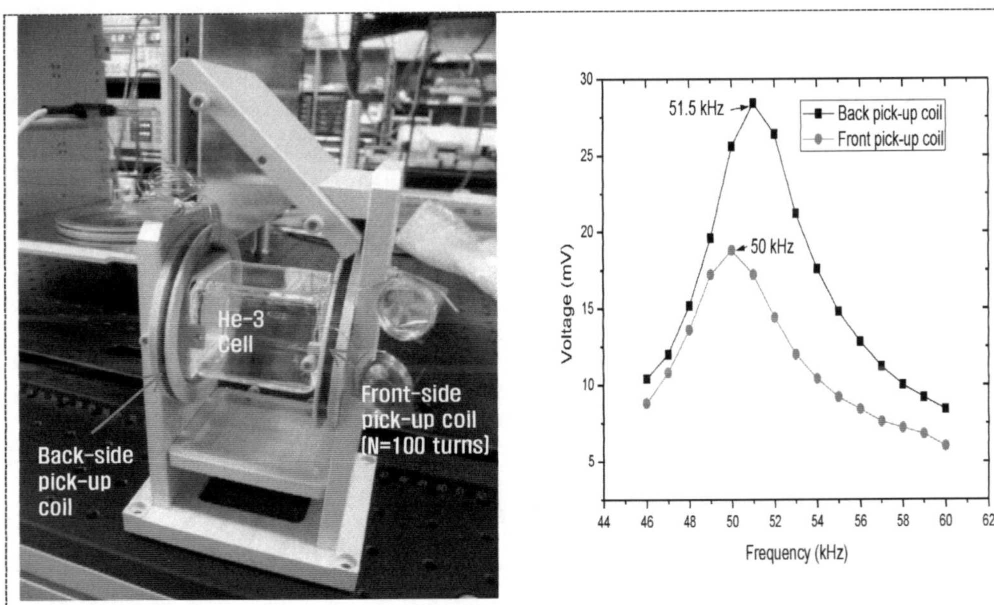

Figure 7. A ring-type pickup coil employed for the NMR signal measurement (left), and the measured resonance frequencies of the LRC circuits for the front- and the back-side pickup coils (right).

Journal of Physics: Conference Series **711** (2016) 012014 doi:10.1088/1742-6596/711/1/012014

4. Conclusions

The characteristics of the magnetostatic cavity and the NMR coils were measured and analyzed experimentally. The developed magnetostatic cavity has a uniform magnetic field distribution and has the advanced capability to compensate a stray environmental magnetic field. These properties of the magnetostatic cavity can led to successful development of a superior ^{3}He polarizer.

A high-purity-cell fabrication station was constructed and used to make a ^{3}He cell for the ^{3}He neutron spin polarizer. High purity ^{3}He cells with various shapes and pressures were fabricated successfully based on Ge-180 glass and fused quartz.

The novel in-situ ^{3}He neutron spin polarizer will be applied in the fields of polarized neutron imaging, TAS, SANS, Reflectometer, etc. Detailed characteristics of the neutron spin polarizer will be reported later.

References

[1] M. Schulz, P. Böni, C. Franz, A. Neubauer, E. Calzada, M. Mühlbauer, B. Schillinger, C. Peiderer, A. Hilger, N.Kardjilov, J. of Phys.: Conference Series **251,** 012068 (2010).

[2] S. Boag, E. Babcock, K. H. Andersen, M. Becker, T. R. Charlton, W. C. Chen, R. M. Dalgliesh, S. D. Elmore, C. D. Frost, T. R. Gentile, R. Lopez Anton, S. R. Parnell, A. K. Petoukhov, M. W. A. Skoda, T. Soldner, Physica B **404**, 2659 (2009).

[3] W. C. Chen, R.Erwin, J. W. McIverIII, S. Watson, C. B. Fu, T. R. Gentile, J. A. Borchers, J. W. Lynn, G. L. Jones, Physica B **404**, 2663 (2009).

[4] E. Babcock, S. Mattauch, A. Ioffe, Nucl. Instr. and Meth. A **625**, 43(2011).

[5] T. Ino, Y. Arimoto, H. M. Shimizu, Y. Sakaguchi, K. Sakai, H. Kira, T. Shinohara, T. Oku, J-I. Suzuki, K. Kakurai, and L-J. Chang, J. of Phys.: Conference Series **340,** 012006 (2012).

Magnetic field optimization and design of a superconducting neutron Wollaston prism

F. Li[1], S. R. Parnell[1, 2, *], T. Wang[1], D. V. Baxter[1] and R. Pynn[1, 3]

[1] Center for Exploration of Energy and Matter, Indiana University, Bloomington, IN, USA

[2] Faculty of Applied Sciences, Delft University of Technology, Mekelweg 15, 2629 JB Delft, The Netherlands

[3] Neutron Sciences Directorate, Oak Ridge National Laboratory, Oak Ridge, TN, USA

E-mail: fankli@indiana.edu

Abstract. We present finite element simulations of a superconducting magnetic Wollaston prism (WP) for neutron scattering with high encoding efficiency and low Larmor phase aberrations. To achieve this, we develop and quantify the design criteria. The validation of simulation tools used for this work are investigated by using two software packages: RADIA and MagNet©. Based on the optimization criteria, various possible configurations of WP are explored with MagNet, from which the best configuration is chosen for further optimization. To optimize the best configuration, the influence of various physical parameters is investigated, including the dimensions, shapes and arrangements of components of the device. The optimum WP was built and measured at both pulsed and constant wavelength neutron sources. In flipping mode, a neutron spin flipping efficiency of ~98.5% was measured independent of neutron wavelength and applied current. In a precession mode, measurements showed a highly linear Larmor phase variation along the horizontal direction with low depolarization. Simulations of the device agree well with the experimental measurements. Possible applications of the device are also discussed.

1. Introduction

In optics, a Wollaston Prism (WP) consisting of two triangular calcite prisms with orthogonal optic axes can split unpolarized light into two spatially separated, linearly-polarized, in-phase beams with orthogonal polarizations. For neutrons, a magnetic WP with an inclined interface separating oppositely directed magnetic fields can produce a spatial separation between the two different neutron spin states [1]. Based on the interference of these two spin states, one can construct a spin echo small angle neutron scattering (SESANS) [2-4] or spin echo modulated small angle neutron scattering (SEMSANS)[5-7] instrument, consisting of four and two such magnetic WPs respectively, which allows the measurement of the density correlations in a scattering object [8]. Classically the change of the neutron beam trajectory due to scattering can be encoded into a net neutron spin Larmor phase by

*The research was performed while S. R. Parnell was at Center for Exploration of Energy and Matter, Indiana University, Bloomington, IN, USA

the magnetic fields [4, 9]. Consequently SESANS resolves the neutron scattering angle independently of the neutron beam collimation and extends the accessible length scale up to several μm [4].

Various types of devices have been employed to implement SESANS, including room temperature WPs [1, 4], permalloy foils [10] and radio frequency (RF) flippers [11]. As discussed in ref. [9], each of these devices has its own shortcomings for measuring long length scales with low aberrations, which requires a high magnetic field intensity to create a large magnetic field integral gradient along the hypotenuse in a single WP. To overcome several of the limitations of the devices described above, we have developed a magnetic WP [9] using high temperature superconducting (HTS) materials. We use HTS tape for current carrying and HTS films for magnetic field confinement. The WP consists of two neighboring triangular field regions sharing the same 45°-inclined HTS film at the common hypotenuse of the two regions. The magnetic field in each region is provided by coils wound on soft iron pole pieces and all the field regions then are surrounded by four HTS films that ensure magnetic field uniformity. The whole assembly is then enclosed by a Mu metal box which has a dual purpose, allowing for flux return and shielding from external fields. The HTS tape and soft iron pole pieces allow high field intensity to be achieved while the HTS films can confine the magnetic flux and thus allow us to create a large, uniform field-integral gradient. The optimized WP was built and measured at both pulsed and constant wavelength neutron sources. In flipping mode, a neutron spin flipping efficiency of ~98.5% was measured independent of neutron wavelength and applied current above 2A at a pulsed neutron source [12] (LENS). In the precession mode, a highly linear Larmor phase variation along the encoding direction with low depolarization was measured using a thin monochromatic neutron beam on the MAGIK instrument [13] at NCNR (NIST Center for Neutron Research). The simulation results of the device agree well with the measurement.

To simulate the device, the validity of two simulation packages was investigated, these were RADIA [14] and MagNet© (Infolytica, Canada). These codes are based on boundary integral and finite element methods respectively. We concluded that RADIA is unable to model superconducting materials with sufficient quality and can produce magnetic field artifacts depending on the meshing size of the components in the region close the superconducting material. Because of its ability to handle the HTS materials involved, MagNet was used in all the simulations and optimization of the device. The criterion for optimization of the WP is introduced below and the critical parameters to build a HTS WP with high encoding efficiency and low phase aberrations are presented. To reduce the thermal load and the risk of quenching the HTS films, a series of simulations are also presented.

2. Model selection

For a neutron traveling in a magnetic field oriented perpendicular to the neutron spin, the spin precesses around the magnetic field vector in a motion called Larmor precession [15]. The accumulated Larmor phase of the neutron spin is proportional to the neutron wavelength and the magnetic field integral along the neutron trajectory.

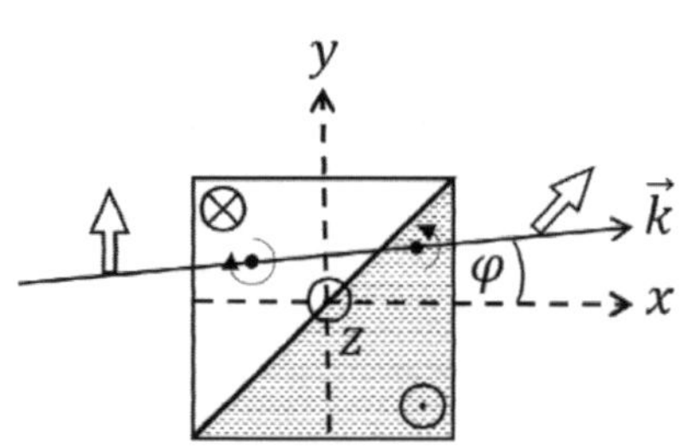

Figure 1 Schematic of a WP where ***k*** is the neutron wave vector and the Larmor precession is indicated. The neutron trajectory has a divergence angle of φ with respect to the x axis in the xy plane. The divergence angle of ψ in the xz plane is not shown.

A single magnetic WP composed of two neighboring triangular field regions is shown in **Error! Reference source not found.** with the Larmor precession indicated. Based on the geometrical symmetry of the device, for a neutron wave vector with divergence angles φ and ψ in the xy plane and the xz plane respectively, the magnetic field integral FI can be expanded to the lowest order as [9].

$$FI = ay + by\varphi + cz\psi + dyz^2 + O(\varphi^2, y^3) \tag{1}$$

where z denotes the direction of the main field and y is the gradient direction of the field integral as shown in **Error! Reference source not found.**, which denotes the encoding direction of the WP. As ref. [9] makes clear, to build a WP with high encoding efficiency and low Larmor phase aberration, the magnitude of a needs to maximized, while minimizing the magnitude of d.

Since most available HTS tapes either contain gadolinium, which is a strong neutron absorber, or are too rigid to wind a well-defined triangular solenoid, the WP is defined by two pairs of gapped solenoids through which neutrons pass. As shown in Figure 2, possible configurations of the WP were simulated using MagNet by manipulating the positions of the HTS films and soft iron pole pieces. For all the configurations, the inclination angle of the boundary between the triangular coils was kept at 45°. The materials used in the simulation were HyMu 80 0.014, Hiperco 50A 0.014 and Copper 5.77e7 Siemens/meter for the Mu metal, pole pieces and coils respectively. The HTS films are defined by a flux tangential boundary condition due to the Meissner effect.

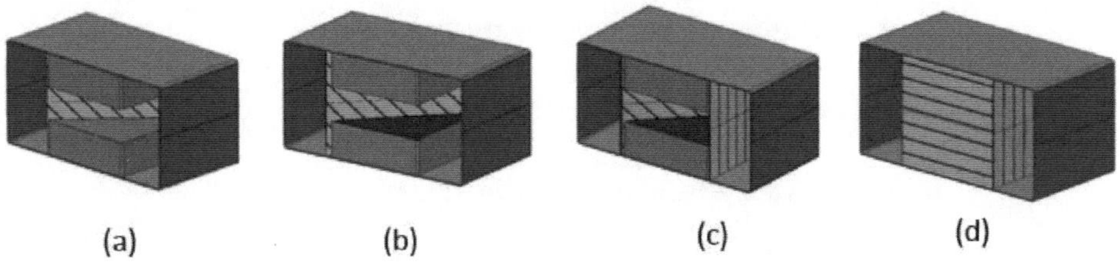

Figure 2. The models we have investigated. Gray with black stripes: HTS film; purple: Mu metal yoke; dark red: soft ion pole pieces; brown: coils wound on pole pieces

Model (a): Hollow triangular coil with HTS film at hypotenuse only

Model (b): Triangular coil filled with pole pieces, HTS film at hypotenuse only

Model (c): Triangular coil filled with pole pieces, HTS films on the side and hypotenuse

Model (d): Triangular coil filled with pole pieces, HTS films on all sides

For each model shown in Figure 2, the corresponding z dependence of the magnetic field is shown in Figure 3, which makes clear that model (*c*) and model (*d*) generate the largest field with the best uniformity. Also the term a and d in the field integral expansion are summarized in Figure 4 showing that model (*d*), which employs soft iron pole pieces and HTS films on all sides, gives the largest linear term a and the lowest aberration term d due to the better field confinement and coupling efficiency between the upper and lower pole pieces. Thus model (*d*) is the best configuration of the WP and is further optimized in the following sections.

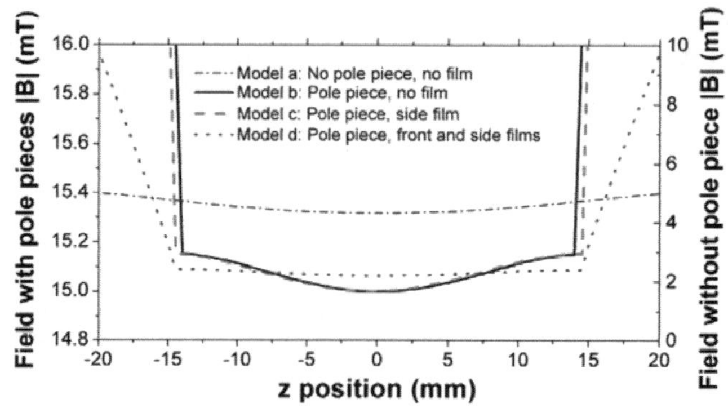

Figure 3. Field intensity of various models along the z direction at the center (y=0) of the triangular-cross section of various models. The gap between the upper and lower coils is 30mm. Model (a) uses the right axis and model (b), (c) and (d) use the left axis.

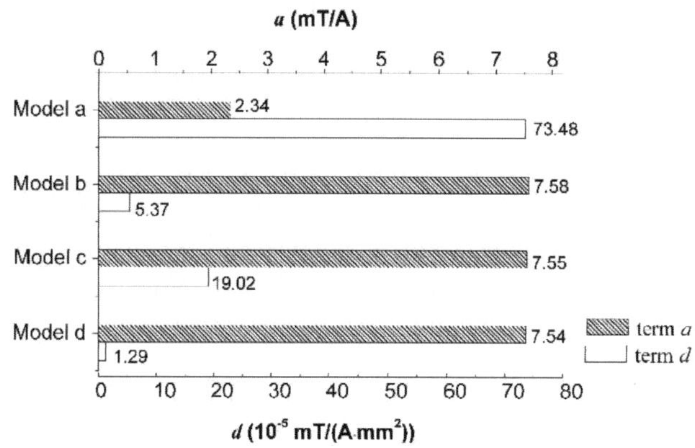

Figure 4. The simulated a and term d for various model configurations. Meshed and blank bars are for term a (upper axis) and d (lower axis) respectively with their values indicated on the right.

3. Optimization of the geometry

In the following section, a WP with soft iron pole pieces and HTS film on each side will be optimized by investigating the dependence of the terms a and d on various physical parameters of the device. Parameters that were varied include the coil height along field direction, coil size, coil arrangement, gap between pole pieces and gaps between coil and HTS films.

3.1. Variation of the coil height

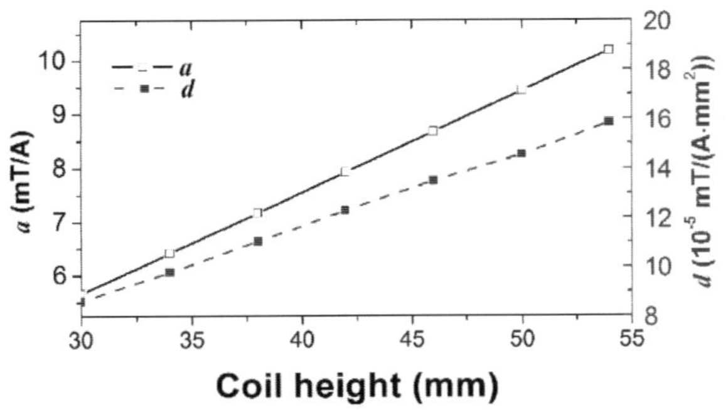

Figure 5. The variation of the terms a and d as a function of coil height. The gap between upper and lower coils is fixed to be 40mm. The current density in the coils was 4 A/mm. Black hollow points are for term a (left axis) and blue solid points are for term d (right axis).

The contribution of the coil height is investigated and shown in Figure 5 and the term a and d both scale linearly with the height of the coils, the ratio of these terms is constant and there is no particular optimum height for the coils. However the maximum height of the coil is limited by the available size of the HTS film.

3.2. Variation of the size and arrangement of coils

Since each triangular-cross section coil has three round corners, they can be configured in at least two possible ways, as shown in Figure 6 (a) and (b). In Figure 6 (a), the two triangular coils are symmetric about the HTS film at the hypotenuse, while for Figure 6 (b), these two triangular coils are configured such that their circumcenters are aligned parallel to the beam direction. The consequent variations of the terms a and d as a function of the size of the coils for both configurations are shown in Figure 6 (c). It is clear that increasing the size of the coil decreases the d term dramatically while the linear term a remains constant. Figure 6 (c) also shows that the arrangement of the two triangles is important, with the triangles arranged as shown in Figure 6 (b) being the optimum arrangement. Again, the maximum size of the coil is mainly restricted by the largest HTS film available from the manufacturer.

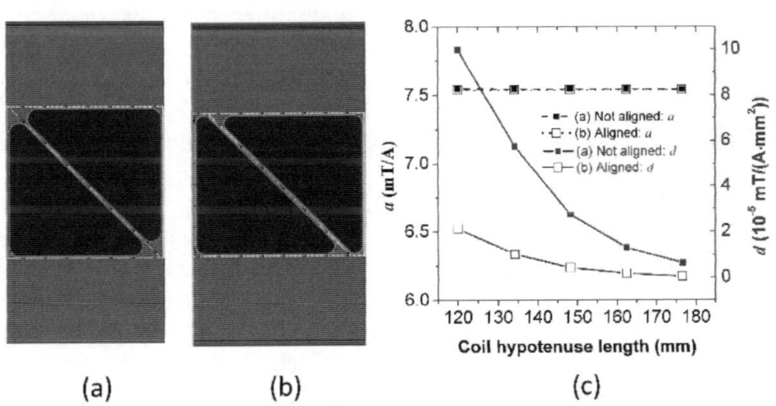

Figure 6. (a) and (b): top view of two possible arrangements of the triangular coils. Purple: Mu metal, brown: pole pieces and black stripe: HTS films. (c): The variation of term a and d term when changing the hypotenuse length of the coil. Solid points are for model (a) while open points are for model (b). The scale for black points is the left axis while the scale for blue points is the right axis.

3.3. Variation of the gap between the pole pieces
To calculate the maximum obtainable beam size, the influence of the gap between the upper and lower coils upon the terms *a* and *d* was simulated. As shown in Figure 7, when increasing the gap, due to the decrease of the coupling efficiency between these two coils, the magnetic field intensity will decrease whilst the field homogeneity increases. Consequently, term *a* will decrease while term *d* will increase, which means the gap should be kept as small as the maximum beam size allows.

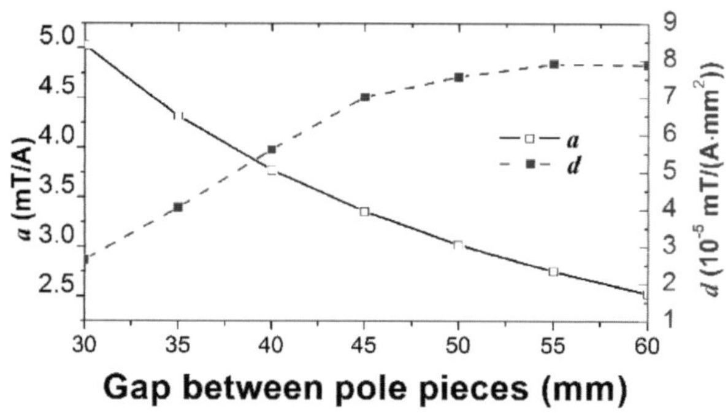

Figure 7. The variation of the terms *a* and *d* as a function of the gap between upper and lower coils. Black hollow points (left axis) are for term *a* and blue solid points (right axis) are for term *d*.

A series of calculations were carried out to determine the effects upon magnetic field-integral homogeneity for gaps between HTS films in the various designs. This was done because it is clearly easier to manufacture designs in which the tolerance on film positions is relaxed. The general finding from these calculations was that gaps between films need to be avoided where possible to confine the magnetic flux.

3.4. Design of the pole pieces
In all the calculations described above, solid triangular-cross-section soft-iron pole pieces with a high permeability were used inside the coils to confine and guide magnetic fields. However these can create an unnecessary thermal load at low temperature and constrain the cooling speed and base temperature. Thus the effect of hollow pole pieces on the field integral expansion was calculated, which shows that a soft iron core with a finite wall thickness is sufficient to guide and confine the magnetic flux between the upper and lower pole pieces provided they are not saturated. The calculation of the field-integral expansion shows no effect on either the term *a* or *d* compared with the model with solid pole pieces.

A key parameter to constrain the maximum current which can be applied is the critical field of the HTS film. Since the edge of the pole piece is close to the HTS films the gradient of the magnetic scalar potential around a sharp corner can be higher than the lower critical magnetic field of the HTS films, which are type-2 superconductors [16]. To minimize the local magnetic field perpendicular to the HTS films and avoid the mixed state in the HTS films, the local field intensity for various types of pole pieces was calculated and is shown in Figure 8, where (b), (c), (d) and (e) demonstrate a corner of the WP indicated by the yellow box in Figure 8 (a). The magnetic field intensity hot spot usually happens where the edge of the pole pieces and coils meet. Figure 8 shows that the minimum local field can only be obtained when the sharp edges on both the HTS coils and the pole pieces can be avoided. The configuration shown in Figure 8 (e) is used in the actual design of the WP.

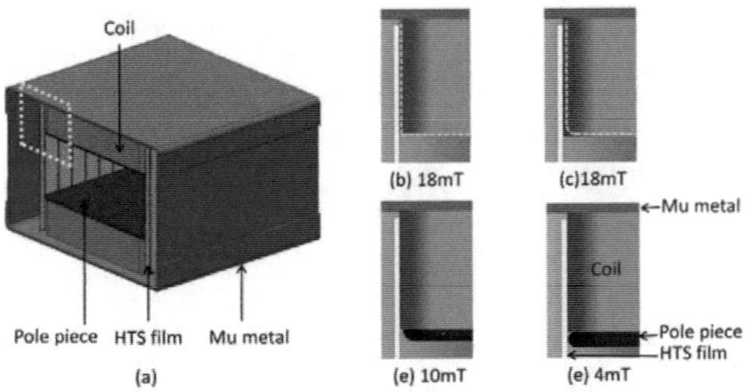

Figure 8. (a) A WP with its front HTS film removed for clarity. (b), (c), (d) and (e) are the possible configurations of the pole piece and coil with respect to the HTS film. Parts (b) – (e) show possible configurations within the yellow dashed square in part (a). The maximum calculated perpendicular component of magnetic fields on the HTS film are indicated below the drawings. Gold, purple, brown and gray indicate the HTS coil, Mu metal, pole piece and HTS film respectively. In (b) and (c), pole pieces are wound to the edge of the coil and are thus invisible so the profile is indicated as yellow dash line. All coils have 20 turns and carry a 30A current.

3.5. Design conclusions

To summarize these simulation results, for a WP with a large linear term a and a small aberration term d, the transverse (y) dimension of the coil needs to be maximized with a particular arrangement while the gap in between the upper and lower coils needs to be no larger than the maximum beam size required. To confine the main magnetic field region and keep the field uniform, gaps between the HTS films need to be avoided where possible. To reduce the thermal mass of the device, hollow pole pieces need to be employed and sharp corners of either the coils or the pole pieces have to be avoided to protect the HTS films from being quenched.

4. Construction of the WP

Based on the optimization criteria described above and the size of the HTS films commercially available [17], the optimized WP configuration is as shown in Figure 9 (a). It consists of 2 pairs of triangular-cross section solenoids, which generate opposite fields on two sides of the shared hypotenuse. All the coils are then yoked in a Mu metal enclosure to ensure magnetic flux return. To reduce the remnant field produced by the soft iron, material with low magnetic coercivity is preferred. Soft iron A1018 is used in the fabrication of the WP due to its availability, although Hiperco 50A 0.014 with lower coercivity is used in the simulation. The complete assembly of the WP is shown in Figure 9 (b). The whole device is sealed inside a helium gas exchange can and can be cooled down to ~30K within 12 hours by a close-cycle refrigerator with a cooling power of 6.7W at a temperature of 20K.

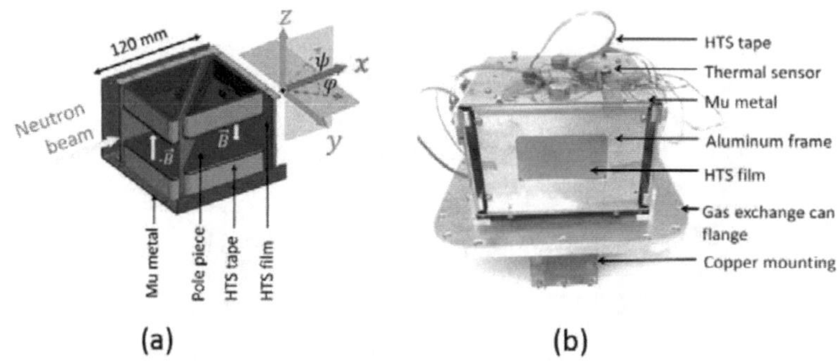

Figure 9. (a) Drawing of the optimum WP. For clarity, one front and one side HTS film have been removed. Similarly, one top and one side Mu metal plate have also been removed. (b) The assembled WP sitting on the flange of the top of the gas exchange can with the outer can removed.

5. Characterization of the WP

The magnetic field integral expansion describes the Larmor precession efficiency of the device, which gives the encoding efficiency and phase aberration of a SESANS instrument. Besides the phase aberration in Larmor precession, the spin transport efficiency of the device is also determined by the magnetic field transitions at the HTS films. For the HTS films, the field transition efficiency is mainly determined by the coupling efficiency between the magnetic fields generated by the device itself and the external guide field. The average angular deviation of the magnetic field inside the device from the z direction is calculated to be 0.004° and 0.014° for $10\times10\text{mm}^2$ and $20\times20\text{mm}^2$ beam sizes respectively. Considering the HTS film at the hypotenuse, again, the calculated average misalignment of the magnetic field vectors from the z direction is 0.004° and 0.01° for $10\times10\text{mm}^2$ and $20\times20\text{mm}^2$ beam sizes respectively. The highly aligned magnetic field vectors contribute to a highly efficient spin transport from external guide field to the WP. The spin transport efficiency of the triangular region is determined by the adiabaticity of the magnetic field [18]. According to the calculation, the minimum adiabaticity parameter is larger than 100, which is >>1 and means the precession can perfectly follow this geometrical rotation of field in the WP and conserve polarization [18].

Various coefficients in the field expansion in equation (1) can be calculated, a=3.045mT/A, b=3.033mT/(A·rad), c=0.045mT/(A·rad) and $d=-1.3\times10^{-5}$ mT /(A·mm^2). As discussed earlier in ref. [9], for a neutron beam of 10Å wavelength with divergence angle ±0.5° in all directions, the integrated polarization efficiency over the whole beam height as a function of spin echo length takes the values shown in Figure 10 for three different beam heights. According to the simulations, a spin echo polarization of 90% can be obtained at a spin echo length of 5μm using a 30mm beam height.

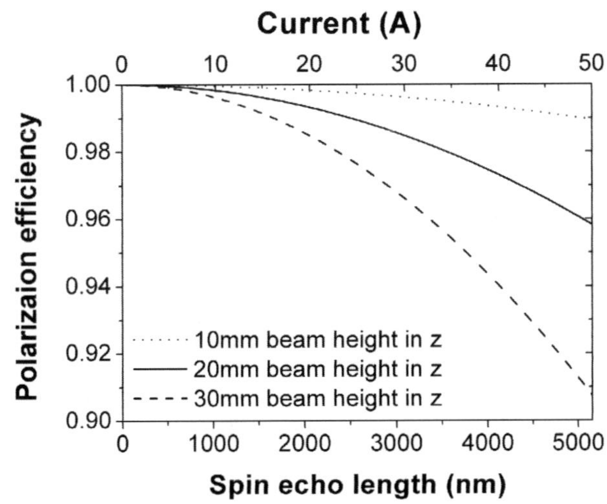

Figure 10. The polarization efficiency integrated over the whole beam height for various beam sizes in the *z* direction as a function of spin echo length. The applied currents for various spin echo lengths are indicated on the top axis. For this calculation, the spin echo length was calculated using a separation distance of 30cm between the WP centers and neutron wavelength of 10Å.

The field intensity of the assembled device was measured using a 1-D hall probe and the performance of the device with polarized neutrons was also measured at both pulsed and constant wavelength neutron sources. All these results are presented in more detail in ref. [9]. In flipping mode, a neutron spin flipping efficiency of ~98.5% was measured independent of neutron wavelength and applied current. In the precession mode, a highly linear variation of the Larmor phase along the horizontal direction was measured with low depolarization. The simulation results of the device agree well with the measurement.

6. Discussion

We have presented our simulations of an HTS magnetic WP. The criteria used for the optimization are introduced and the contributions of the critical parameters of the WP for a high encoding efficiency with low phase aberrations are evaluated. As an example, the expected performance of a SESANS instrument is presented with a high polarization at long spin echo length. The simulations using MagNet© indicate its capability for defining the HTS materials. Also our neutron measurements of the WP make it clear that the HTS films can be employed to split two neighboring magnetic field regions with high spin transport efficiency. Thus both the simulation package and the HTS films may have potential applications to facilitate the design of neutron spin-manipulation devices.

To further simplify the maintenance of the device and extend the achievable magnetic field intensity, a second generation of the device has been constructed with optimized thermal and electrical conductivity, which can possibly allow at least 50mT to be achieved without the gas exchange can, and decreases the coupling distance to the external guide field. Thus, a larger encoding efficiency with higher magnetic coupling efficiency can be achieved.

Besides the application in SESANS, the separated two spin states can record the phase change introduced by a scattering object, thus the gradient of the phase distribution of the object along the hypotenuse of the device can be measured by the interference of these two spin states, which can be used to implement neutron differential interference contrast (DIC) radiography. The same setup of the DIC radiography can also be employed to do spin-echo modulation small angle neutron scattering. The implementation of SEMSANS using two of such WPs has been achieved at a reactor source[7].

Recently we also proposed a new neutron spin echo method to measure the phonon excitation life time using two of such WPs with a rectangular magnetic field region between them on both the incident and scattered neutron beams on a triple-axis spectrometer [19]. By introducing well-defined magnetic field regions with triangular and rectangular cross sections, the lines of constant spin echo phase can be tuned electromagnetically to a wide range of phonon energies and group velocities simply by changing currents. By introducing electromagnetic tuning and allowing higher current to be applied, the new method may accommodate a wider range of phonon energies and group velocities than the traditional resonant spin echo technique.

7. Acknowledgement

A grant from the National Science Foundation (grant number DMR-0956741) supported design and construction of our first neutron spin manipulation devices (a spin flipper and a device for spherical neutron polarimetry) as well as conceptual design of the WP described here. Construction of the prototype WP was supported by the STTR program of the US Department of Energy (grant number DE-SC0009584).

References

[1] Pynn R, Fitzsimmons M R, Lee W T, Stonaha P, Shah V R, Washington A L, Kirby B J, Majkrzak C F and Maranville B B 2009 *Physica B* **404** 2582-2584
[2] Uca O, Spin-Echo Small-Angle Neutron Scattering Development, in, Delft University Press, 2003.
[3] Rekveldt M T, Plomp J, Bouwman W G, Kraan W H, Grigoriev S and Blaauw M 2005 *Rev. Sci. Instrum.* **76** 033901
[4] Pynn R, Lee W T, Stonaha P, Shah V R, Washington A L, Kirby B J, Majkrzak C F and Maranville B B 2008 *Rev. Sci. Instrum.* **79** 063901
[5] Strobl M, Tremsin A S, Hilger A, Wieder F, Kardjilov N, Manke I, Bouwman W G and Plomp J 2012 *J. Appl. Phys.* **112** 014503
[6] Strobl M, Wieder F, Duif C P, Hilger A, Kardjilov N, Manke I and Bouwman W G 2012 *Physica B* **407** 4132-4135
[7] Li F, Parnell S R, Bai H, Yang W, Hamilton W A, Maranville B B, Ashkar R, Baxter D V, Cremer J T and Pynn R 2016 *J. Appl. Crystallogr.* **49** 55-63
[8] Andersson R, van Heijkamp L F, de Schepper I M and Bouwman W G 2008 *J. Appl. Crystallogr.* **41** 868-885
[9] Li F, Parnell S R, Hamilton W A, Maranville B B, Wang T, Semerad R, Baxter D V, Cremer J T and Pynn R 2014 *Rev. Sci. Instrum.* **85** 053303
[10] Rekveldt M T, Duif C P, Kraan W H, Plomp J and Bouwman W G 2008 *Rev. Sci. Instrum.* **79** 015113
[11] Plomp J, Spin-echo development for a time of flight neutron reflectometer, in, Delft University of Technology, South Holland, Netherlands, 2009.
[12] Baxter D V, Cameron J M, Derenchuk V P, Lavelle C M, Leuschner M B, Lone M A, Meyer H O, Rinckel T and Snow W M 2005 *Nucl. Instrum. Methods Phys. Res., Sect. B* **241** 209-212
[13] Dura J A, Pierce D J, Majkrzak C F, Maliszewskyj N C, McGillivray D J, Lösche M, O'Donovan K V, Mihailescu M, Perez-Salas U, Worcester D L and White S H 2006 *Rev. Sci. Instrum.* **77** 074301
[14] Elleaume P, Chubar O and Chavanne J, Computing 3D magnetic fields from insertion devices, in: Particle Accelerator Conference, 1997. Proceedings of the 1997, 1997, pp. 3509-3511 vol.3503.
[15] Bloch F 1946 *Phys. Rev.* **70** 460-474
[16] Böhmer C, Brandstätter G and Weber H W 1997 *Supercond. Sci. Technol.* **10** A1

[17] ceraco ceramic coating GmbH *http://www.ceraco.de/hts-films/*
[18] Seeger P A and Daemen L L 2001 *Nucl. Instrum. Methods Phys. Res., Sect. A* **457** 338-346
[19] Li F and Pynn R 2014 *J. Appl. Crystallogr.* **47** 1849-1854

Interaction of neutrons with a birefringent medium moving with an acceleration

A.I. Frank

Frank Laboratory of Neutron Physics, Joint Institute for Nuclear Research, Dubna, Russia

frank@nf.jinr.ru

Abstract. Recent experiments demonstrated that the energy of a neutron traversing an accelerated sample of a refractive medium changes. Later, it was realized that such an accelerated-medium effect (AME) is quite a general phenomenon. This paper discusses some special features of the effect for a birefringent medium. In this case, AME shows quite new features. In neutron optics, where birefringence is due to the spin dependence of the refractive index, AME results in a nonstationary state with a precessing spin.

1. Introduction

The first theoretical study devoted to the neutron wave interaction with moving matter appeared in the late 1970s [1]. It triggered the appearance of a number of experimental [3–5] and theoretical [6,7] studies. A review of those studies and a comparison of the results obtained in conventional and neutron optics are given in [8]. By analogy with light optics, neutron experiments with moving media became known as neutron Fizeau experiments.

While having a number of similar features, phenomena accompanying the propagation of electromagnetic and neutron waves through moving matter are markedly different [9]. Fresnel's drag of light and the phase shift of electromagnetic waves in the Fizeau experiment are due to the relativistic nature of the velocity transformation. In contrast, neutron optics deals with nonrelativistic particles, so that in order to go over to a different reference frame, one has to perform a Galilean transformation of the respective wave function. Further, neutron interaction with matter can be described in terms of an effective potential,

$$U = \frac{2\pi\hbar^2}{m}\rho b\,, \tag{1}$$

where ρ is the spatial density of nuclei, and b is the volume-averaged coherent scattering length. The motion of matter itself leaving the effective potential unchanged does not have any effects on the neutron and the phase shift can only be obtained by the motion of sample boundaries. If, however, a layer of matter moves together with its boundaries as a coherent whole, the wave frequency in matter as measured in the laboratory frame is different from that in a vacuum. This will be discussed in more detail in the next section.

In the case of the motion of a matter layer at a constant velocity, Doppler frequency shifts resulting from the passage of a wave through two boundaries of the sample are equal in magnitude but opposite in sign. The net effect is zero, and the frequency of the wave that passed through the moving sample does not change. In the case of an arbitrary motion, such a cancellation does not occur, so that the frequency of the wave transmitted through the sample differs from that of the incident wave. For the first time, this was indicated by Tanaka in [10].

Apparently, the problem of neutrons passing through a layer of matter moving with a linear acceleration was first considered by Kowalski [11]. His theoretical approach was based on the notion of the neutron group and phase velocities and on the calculation of the time of wave propagation between relevant points in stationary and moving reference frames. As an intermediate result, he arrived at the conclusion that the energy of a neutron changes after its transmission through a sample moving at a not-too-high acceleration. Later on, the same issue was considered by Nosov and Frank [12]. Their analysis, in which a consistent calculation of the velocity of a neutron upon entering the sample and exiting it through another surface was performed, was in fact based on the classical approach.

The first brief report on the experimental observation of the change in the energy of neutrons propagating through an accelerated layer of matter appeared in [13]. The results of a more detailed study of this effect, along with basic theoretical conclusions, were presented in [14]. In the same paper, the authors put forth the idea that the observed effect was a manifestation of quite a general phenomenon – the accelerated medium effect (AME) – inherent in waves and particles of different nature. Recently, the acceleration and deceleration of neutrons were observed by applying a specific time-of-flight method [15]. The acceleration of the samples in those experiments was several tens of g units, the value of the energy transferred to a neutron fell within the range of $(2\div6)\times10^{-10}$ eV, and the experimental results obtained by two different methods were in agreement with the theory to a precision of about 10%. Recently the energy change of neutrons passing through the accelerated crystal close to the Bragg angle was observed in [16]. The first stage of AME investigations with UCN was summarized in [17], where some features of neutron interaction with an accelerated sample of birefringent matter were also briefly discussed. The similar phenomena in neutron and neutrino physics were also discussed in [18].

In the next section, we discuss the problem of neutron wave propagation through a moving medium and show how the propagation of a neutron wave through a layer of conventional refracting matter moving with an acceleration leads to a change in energy. In the following two sections we discuss some features of the AME manifestation in the case of a birefringent medium and some possibilities of their experimental observation.

2. Doppler effect upon the refraction of a neutron wave at the boundary of a moving medium and accelerated matter effect

We now consider the problem of neutron-wave refraction at the boundary of a half-infinite layer of matter moving at a constant velocity V along the x axis. Representing, as usual, the initial state of the neutron as a plane wave, $\psi(x, t) = \exp[i\,(k_0x - \omega_0t)]$, we find the neutron wave function in the reference frame co-moving with matter. According to [19] it has the form

$$\psi'(x',t) = \psi(x'+Vt,t)\,exp\left[-i\left(k_Vx'+\omega_Vt\right)\right], \tag{2}$$

where

$$k_V = \frac{mV}{\hbar}, \quad \omega_V = \frac{mV^2}{2\hbar}. \tag{3}$$

In this reference frame, the wave number and frequency of the incident wave are

$$k' = k_0 - k_V, \quad \omega' = \omega_0 + \omega_v - kV\,. \tag{4}$$

Upon refraction at the matter boundary, which is stationary in this reference frame, only the wave number changes, so that the wave function inside matter assumes the form

$$\psi_i'(x',t) = exp\left[i\left(k'n'x' - \omega't\right)\right]. \tag{5}$$

where $n' = n(k')$ is the wave-number-dependent refractive index,

$$n(k) = \sqrt{1 - k_b^2/k^2}\,, \tag{6}$$

and $k_b^2 = 4\pi\rho b$ is the boundary value of the wave number.

After substituting (4) into (5) and returning to the laboratory frame we find for the wave number in the moving matter that

$$k_i = n'k_0\left(1+\frac{1-n'}{n'}\frac{V}{v_0}\right) , \tag{7}$$

where v_0 is the neutron velocity in a vacuum. One can easily see that this formula coincides with the well-known expression for the Fresnel drag in a moving medium (see, for example [20]).

For the frequency of a neutron wave in a medium, as measured in the laboratory frame, we obtain

$$\omega_i = \omega_0 + (n'-1)(k_0 - k_V)V .$$

Assuming that velocity of matter is small in relation to v_0, we obtain the following expression to the first order terms in V/v_0 :

$$\omega_i = \omega_0 + (n'-1)k_0 V \qquad (V << v_0). \tag{8}$$

Equation (8) describes the Doppler effect occurring upon neutron-wave refraction at the boundary of a moving medium. It coincides with the similar expression obtained in [21] to the first order in V/c for light refraction at the boundary of a moving medium.

We now consider the propagation of a wave through a moving sample and find the phase shift caused by the presence of matter. For a stationary sample, it is $\chi = (k_i - k_0)d$, where d is the sample thickness. Substituting expression (9) into it and setting $k_i = n'k_0 - (n'-1)k_V$, we obtain

$$\chi = (n'-1)(k_0 - k_V)d . \tag{9}$$

Substituting into (9) the definition $n' = \sqrt{(k_0-k_V)^2 - k_b^2} \times (k_0 - k_v)^{-1}$ we arrive at the relation

$$\chi = \left[\sqrt{(k_0-k_V)^2 - k_b^2} - (k_0 - k_V)\right]d , \tag{10}$$

which coincides with formula (5) from [6].

As concerns the frequency of a wave passing through a sample, it was indicated in the Introduction that in the case where a layer of matter moves at a constant velocity, the frequency shifts resulting from the passage of the wave through two boundaries are equal in magnitude but opposite in sign. The net effect is then zero. In the case of an accelerated motion, the frequency shifts at the entrance and exit surfaces of the sample are unequal and do not cancel each other, since, during the time of wave propagation through the sample, the velocity of the boundaries changes by $\Delta V = a\tau$, where a is the acceleration of the sample and τ is the propagation time. Setting the latter to $\tau = d/(nv_0)$ and neglecting, for the time being, the dispersion of the medium, we find that the difference between the frequencies of the incident and transmitted waves is $\Delta\omega = ad(1-n)k_0/nv_0$ The change in the energy is then given by

$$\Delta E = mad\left(\frac{1-n}{n}\right) \qquad (a\tau << v_0), \tag{11}$$

in perfect agreement with the results obtained in [11, 12]. Relation (11) was obtained to the first order in V/v_0 and $\Delta V/v_0$.

It should be noted that the above way of deriving (11) for AME is only valid under the adopted assumptions, since it completely ignores the fact of the accelerated motion of the medium itself. In neutron optics, this does not result in an error, at least at moderate accelerations, since (and as long as) the description of the medium with the aid of an effective potential (1) is true here and the motion of the region within which the potential is operative by no means affects the state of the particle inside it.

At the same time some arguments were presented recently that the concept of effective potential may be wrong in the case of a very large acceleration of matter [22].

It is worth mentioning that for derivation of Eq. (11) we have only used the fact that the wave number in accelerated matter is not equal to that in a vacuum. Thus, the effect should occur at an accelerated motion of refractive matter as well as of a finite spatial region containing a force field. Below we do not draw a distinction between these two cases but define the refractive index as the ratio of the wave number in a medium (or in the area of the action of the potential) and in a vacuum, $n = k/k_0$. It allows us to use formula (11) in both cases. The refractive index for an area with the potential U is given by

$$n = \sqrt{1 - \frac{U}{E}}, \tag{12}$$

where E is the neutron energy in free space $(E > U)$.

3. Interaction of neutrons with moving matter in the case of double refraction

3.1 The case of permanent velocity

Birefrigent matter is characterized by two refractive indices in accordance with the two values of polarization of the incident wave. In neutron optics, two values of the refractive index, $n_\pm$, correspond to two different projections of the neutron spin onto the physical axis. For a sample moving with a constant speed, relations (7) and (9) hold. In the case of arbitrary polarization of the initial wave function

$$\Psi_0(x,t) = \begin{pmatrix} A_+^0 \\ A_-^0 \end{pmatrix} exp\left[i(kx - \omega t)\right],$$

the state inside the matter is

$$\Psi_i(x,t) = \begin{pmatrix} A_+^0\, exp(ik_{i+}x) \\ A_-^0\, exp(ik_{i-}x) \end{pmatrix} exp(-i\omega_i t), \tag{13}$$

and the spin is rotating during the time of neutron propagation inside the sample. After passing the moving sample the state is

$$\Psi_f(x,t) = \begin{pmatrix} A_+^0\, exp\, i\chi_+ \\ A_-^0\, exp\, i\chi_- \end{pmatrix} exp\left[i(kx - \omega t)\right], \tag{14}$$

where the phase shifts $\chi_\pm$ caused by the presence of the sample are:

$$\chi_\pm = (n'_\pm - 1)(k_0 - k_V)d\,. \tag{15}$$

The phase difference $\Phi(V) = \chi_+(V) - \chi_-(V)$ determines the spin rotation angle [23].

3.2 The case of accelerated matter

As it has been mentioned above, the neutron wave number changes only at the matter boundaries. So, if we talk about the effects arising prior to refraction at the exit surface, the motion of the sample affects only the effective path length travelled by the neutron inside the sample. In the case of an accelerated motion, the effective thickness of the sample is apparently equal to $d_{eff} = d + a\tau^2/2$.

Let us now address the issue concerning the frequencies of the two waves transmitted through accelerated matter. It is obvious that in the case of double refraction we have to modify Eq. (11) as

$$\Delta E_{\pm} = mad\left(\frac{1-n_{\pm}}{n_{\pm}}\right), \qquad \Delta\omega_{\pm} = \frac{\Delta E_{\pm}}{\hbar}. \tag{16}$$

After the passage of an accelerated doubly refracting sample, the two spin components of the neutron wave function differ in frequency, so that the wave function is a time-dependent superposition and the final state has now the form

$$\Psi(x,t) = \begin{pmatrix} A_+ \\ A_- \end{pmatrix} exp\left[i(kx-\omega t)\right], \quad A_{\pm} = A_{\pm}^0\, exp\left[-i\left(\Delta k_{\pm}x + \Delta\omega_{\pm}t + \chi_{\pm}\right)\right]. \tag{17}$$

The wave function $\Psi(x,t)$ describes a nonstationary state with a precessing spin. The precession angle is obtained as the time-dependent difference of the phase angles of the two spin components; that is,

$$\phi(x,t) = \left(k_+ - k_-\right)x + \Omega t + \Phi(t), \tag{18}$$

where $k_{\pm} = k_0 + \Delta k_{\pm}$, $\Omega = \omega_+ - \omega_-$ and $k_+^2 - k_-^2 = 2m\Omega/\hbar$. Introducing the average velocity $\hat{v}$ for the two components, we arrive at an expression for the spin precession angle in the form

$$\phi(x,t) = \Omega\left(t - \frac{x}{\hat{v}}\right) + \Phi(t), \quad \hat{v} = \frac{\hbar}{m}\frac{k_+ + k_-}{2}. \tag{19}$$

A time dependence arises in the phase shift Φ since, in the accelerated-motion case, which is considered here, the velocity V and the wave number kV are not constants; as a result, the phases $\chi_{\pm}$ given by Eqs. (9) and (10) change with time. In the approximation of a small change in energy, $\Delta E_{\pm}/E << 1$, the effective velocity $\hat{v}$ coincides with the classical velocity. As follows from Eq. (19), the direction of the neutron spin is invariable in the reference frame moving at the neutron velocity $\hat{v}$ with respect to the laboratory frame. At a fixed observation point, however, the spin direction changes with time periodically at the frequency Ω. The situation is similar to that which was first observed in the experiment reported in [24], where the energy of the two spin components was changed by means of two resonant spin flippers operating at slightly different frequencies. It is also quite typical of neutron resonant spin-echo instruments and related facilities [25-27]. A similar situation must occur upon the dynamic reflection and refraction of neutrons at the boundary of a magnet having a variable induction [28, 29].

In all of the aforementioned cases [24–29], a nonstationary state arises owing to the interaction of the neutron magnetic moment with an alternating magnetic field – that is, via a strictly quantum process. A feature distinguishing the AME case for birefringent matter under discussion is that a nonstationary state arises as a difference effect in a classical phenomenon such as the Doppler frequency shift.

From the practical point of view, it is very important that the periodic change under discussion in the polarization direction can be measured even if the oscillation frequency Ω and the respective energy transfer $\hbar\Omega$ are quite small. Under the assumption that $\Delta n = n_+ - n_- \ll n = \left(n_+ + n_-\right)/2$ it follows from (16) that

$$|\Omega| = |\Delta\omega_+ - \Delta\omega_-| \simeq \frac{mad}{\hbar}\frac{\Delta n}{n}. \tag{20}$$

3.3 Birefringence in neutron optics and AME

In neutron optics, several physical reasons may be responsible for the origin of birefringence. First, we will dwell on the simplest case of interaction between the neutron magnetic moment μ and the magnetic field B. In accordance with (12), the spatial domain where a nonzero magnetic field is present is characterized by two values of the refractive index:

$$n_{\pm} = \sqrt{1 \mp \frac{\mu B}{E}}. \tag{21}$$

In line with the aforesaid, neutron propagation through such a domain moving with an acceleration results in the formation of a nonstationary state. From Eqs. (16) and (20), one can readily find that in a conventional weak-field approximation $\mu B << E$, the beat frequency Ω is

$$\Omega \simeq \frac{\mu B}{\hbar}\frac{mad}{E} = \omega_L \frac{mad}{2E}, \tag{22}$$

where $\omega_L = 2\mu B/\hbar$ is the Larmor frequency.

It is highly probable that this simplest case is the most attractive for a first experimental demonstration of the neutron spin rotation resulting from the interaction with an accelerated birefringent medium. Below, we will discuss some details of such an experiment.

In addition to a moving domain where a magnetic field is present, it is useful to consider the case of a magnetic sample. If the sample material is characterized by an effective potential U and a magnetic induction B, the two values of the neutron refractive index are

$$n_{\pm} = \sqrt{1 - \frac{U \mp \mu B}{E}}. \tag{23}$$

If $U \pm \mu B << E$, relation (22) remains valid in this case inclusive.

A nonmagnetic sample placed in a magnetic field can also act as a birefringent material for a neutron wave [30]. In a sense, we might speak here about ordinary matter placed in a birefringent medium with the refractive index given by (21). The two spin components then have the wave numbers

$$k_{\pm} = k_0\sqrt{1 \mp \frac{\mu B}{E}}, \quad E = \frac{\hbar^2}{2m}k_0^2, \tag{24}$$

where k_0 is the wave number in the absence of a field. The effective refractive index of the sample is then

$$n_{\pm} = \sqrt{1 - \frac{U}{E \mp \mu B}}. \tag{25}$$

The accelerated motion of a sample in a constant magnetic field also leads to the formation of states with unequal frequencies for the two spin components of the wave function and to the precession frequency caused by AME and given by

$$\Omega \simeq \omega_L \frac{U}{2E} maL, \quad \left(\frac{\mu B}{E} << 1\right). \tag{26}$$

In general, birefringence may also occur in the absence of a magnetic field. By way of example, we focus on the nuclear pseudomagnetism phenomenon occurring in the case where polarized neutrons propagate through matter containing polarized nuclei [31, 32]. Because of the spin dependence of nuclear forces, the coherence lengths for neutron–nucleus scattering, $b_{\pm}$, are different for the two values of the total spin of the neutron and nucleus, $J = I \pm 1/2$, where I is the spin of the nucleus. Employing the notation $b = (b_+ + b_-)/2$ and $\Delta b = b_+ - b_-$, we obtain an expression for the refractive indices in the form

$$n_{\pm} = \sqrt{n^2 \mp p_{nuc} \frac{2\pi\rho}{k_0^2} \Delta b}\ , \tag{27}$$

where p_{nuc} is the polarization of the sample and $n^2 = 1 - 4\pi\rho b/k_0^2$. Assuming that the effect of nuclear pseudomagnetism may be detected for small values of polarization, we consider the case of p_{nuc}<<1. Then for any realistic values of k_0, relation (29) can be written as

$$n_{\pm} = n \mp p_{nuc} \frac{\pi\rho}{nk_0^2} \Delta b\ . \tag{28}$$

Substituting (28) into (20), we obtain a relation for the frequency of precession arising due to AME:

$$\Omega = p_{nuc} \frac{mad}{\hbar} \frac{2\pi\rho}{n^2 k_0^2} \Delta b\ . \tag{29}$$

The inversely proportional dependence of the precession frequency on the square of the wave number gives sufficient grounds to believe that the spin-dependent scattering lengths Δb can be measured by using very slow neutrons even in the case of rather low nuclear polarization.

Finally, birefringence could be due to the parity violation in neutron-nucleus interactions. But AME in this case is very small [18] and we shall not discuss it here.

3.4 Possibility of experimental observation of AME in the case of birefringence

Spin rotation in the case of neutron propagation through an accelerated birefringent sample can be detected in experiments similar to those where one observes the Larmor spin precession in a refracting sample [33]. The principal difference is that in this case one has to measure the time dependence of the spin precession angle within the time that is commensurate with the spin-rotation period $T \approx 1/\Omega$. This means that the total rotation angle $\phi_{max} = \Omega t$ can be comparatively large even at a very small value of the ratio $\Delta n/n$. On the other hand, the measurement time is limited by the time interval within which the accelerated sample remains inside the measuring instrument used. It is obvious that in order to improve statistics, the measurement must be repeated many times.

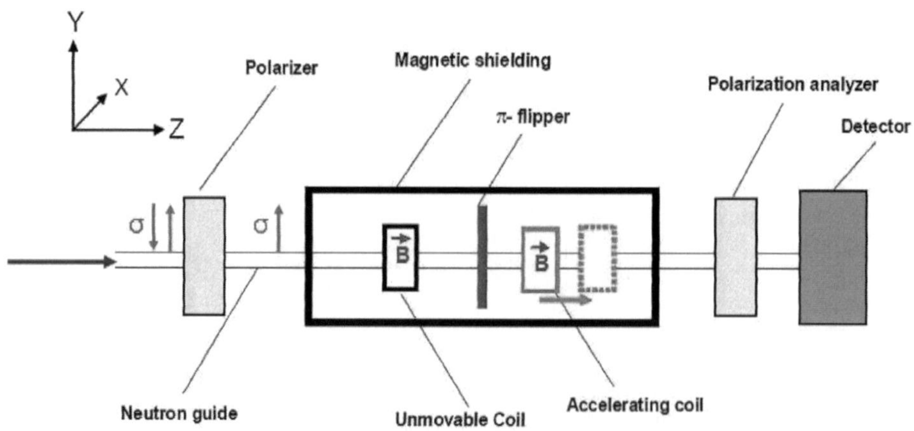

Figure 1. Layout of an experiment aimed at observing the rotation of the spin of a neutron propagating through a region with a magnetic field and moving with an acceleration.

As an example, we will consider a possible experiment aimed at observing AME in the case where a bounded region with a magnetic field moves with an acceleration (see Fig. 1). The respective experimental facility is a classic neutron spin-echo spectrometer [34]. Neutrons polarized along the y axis enter nonadiabatically the region where there exists a magnetic field *B* directed along the *z* axis and created by an immobile coil. Throughout the time within which a neutron remains inside the coil, its spin precesses in an ordinary fashion. Having passed through a π-flipper, neutrons enter the second coil identical to the first one. In the stationary case, the spin-precession angles in the two coils cancel each other exactly, so that the total precession angle is zero. According to (19), the accelerated motion of the second coil along the beam direction leads to the appearance of an additional time-dependent precession angle.

Since the transmission of an analyzer placed in front of the detector depends on the spin direction, the polarization-vector rotation associated with the motion of the second coil leads to a change in the detector counting rate. If the effect is small, an additional π/2-flipper can be placed in front of the analyzer. This corresponds to the well-known method of orthogonally crossed polarizer–analyzer pairs.

Figure 2, which illustrates the estimated results of such an experiment, gives the spin-precession angle φ and the quantity $1-\cos\phi$ versus time. The latter is proportional to the detector counting rate under ideal conditions. By ideal conditions, we mean a hundred-percent efficiency of the polarizer and analyzer, the total monochromation of the beam, and the absence of any background. The instant at which the coil begins its motion is taken for a reference point. The measurement starts at a later instant, whereby one takes into account the neutron time of flight from the coil to the detector (0.4 s).

These calculations were performed for an experiment with ultracold neutrons of speed *4.5* m/s. We assumed that the magnetic field is 100 μT in both coils and that the path length inside the coil is 1 cm. Under these conditions, the neutron refractive indices (21) for the two spin components are estimated at $n_{\pm} = 1 \mp 2.8\times10^{-5}$.

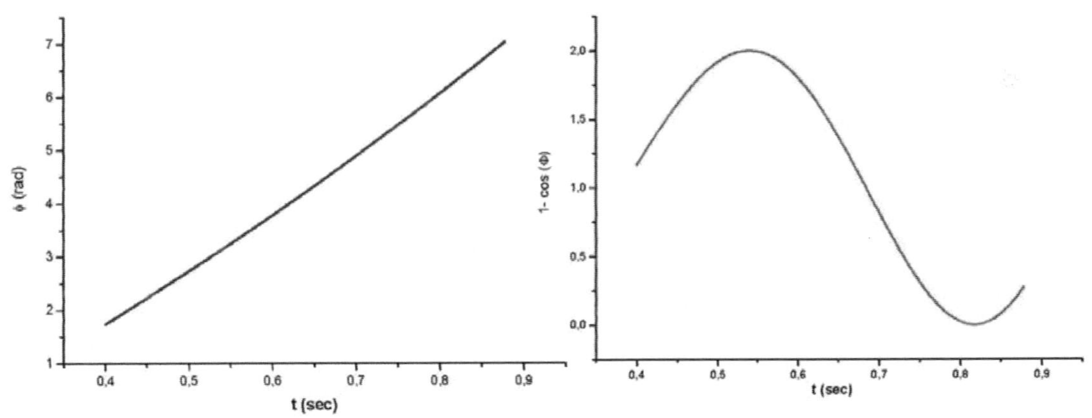

Figure 2. Results obtained by simulating the experiment illustrated in Fig. 1: (a) spin rotation angle as a function of time and (b) variation of the detector counting rate under ideal conditions as a function of time.

Relation (22) allows us to estimate the precession frequency at $\Omega \approx 4.2$ rad/s. The corresponding energy transfer to the neutron is $\Delta E = \pm\hbar\Omega/2 \approx \pm1.5\times10^{-15}$ eV. In the calculations, we assume that the coil moves at an acceleration of about 50 cm/s^2 over the period of $T = 0.5$ s, traveling within this period a distance of about 10 cm and reaching a speed of about 24.5 cm/s. The detector is placed at a distance of 180 cm from the initial position of the coil.

The results of the calculation are likely to favor the statement that for the above properties of a birefringent sample and the adopted parameters of its motion, the AME-induced effect of a nonstationary spin rotation can be reliably detected. Specifically, Fig. 2 shows that the angle of polarization-vector rotation is close to 2π, whereas the sensitivity of modern experiments permits measuring values that are many orders of magnitude smaller.

4. Conclusion

This article has been devoted to the discussion of a new aspect of the AME – namely, the case of an accelerated motion of birefringent matter. In this case, which was mentioned earlier in [17], the AME acquires quite new features. In neutron optics, birefringence is caused by the spin dependence of the refractive index; as a result, the AME gives rise to a nonstationary state with a rotating spin.

Physical reasons that lead to the double-refraction phenomenon in neutron optics have been considered, and the respective formulas for the resulting spin-precession frequency have been derived for this case. For the simplest case of double refraction in a magnetic field, we have considered a possible implementation of an experiment aimed at observing the effect under discussion. The results of a simulation of an experiment that would involve ultracold neutrons and in which a magnetic field with the refractive indices of $n_{\pm} = 1 \pm 3\times10^{-5}$ would move at an acceleration of 0.05 g for 0.5 s have been presented. Under such conditions, the energy transfer to the neutron would be about $\pm1.5\times10^{-15}$ eV. Nevertheless, the AME induced by the precession of the neutron spin is quite measurable.

5. Acknowledgements

This work was partly supported by the RFBR grant No.11-02-00271-a. The author is grateful to V.G. Kulin, D.V. Kustov and V.A. Naumov for fruitful discussions.

References

[1] Horn M A and Zeilinger A 1979 *Neutron Interferometry*, Eds. Bonse U and Rauch H (Clarendon Press, Oxford) 350
[2] Klein A G, Opat G I, Cimmino A, Zeilinger A, Trimer W and Gähler R 1981 *Phys. Rev. Lett.* **46** 1551
[3] Arif M, Kaiser H, Werner S, Cimmino A A, Hamilton W A, Klein A G and Opat G I 1985 *Phys. Rev.* **A 31** 1203
[4] Arif M, Kaiser H, Clothier R., Werner S A , Hamilton W A, Cimmino A and Klein A G. 1989 *Phys. Rev.* **A 39** 931
[5] Bonse U and Rump A 1986 *Phys. Rev. Lett.* **56** 2441
[6] Horne M A, Zeilinger A, Klein A G and Opat G I, 1983 *Phys. Rev.A* **28** 1
[7] Bonse U and Rump A 1988 *Phys.Rev. A* **37** 1059
[8] Neutze R and Stedman G E 1998 *Phys. Rev. A*, **58**, 82
[9] A. Peres 1983 *Am. J. Phys.* **51** 947
[10] Tanaka K, 1982 *Phys. Rev. A* **25**, 385
[11] Kowalski F V 1993 *Phys. Lett. A* **182**, 335
[12] Nosov V G and Frank A I 1998 *Phys. At. Nucl.* **61** 613
[13] Frank A I, Geltenbort P, Kulin G V, Kustov D V, Nosov V G and Strepetov A N 2006 *JETP Letters*, **84** 363

[14] Frank A I, Geltenbort P, Jentschel M, Kustov D V, Kulin G V, Nosov V G and Strepetov A N 2008 *Phys. At. Nucl.* **71** 1656
[15] Frank A I, Geltenbort P, Jentschel M, Kulin G V, Kustov D V, Nosov V G and Strepetov A N 2011 *JETP Letters* **93** 361
[16] Voronin V V, Berdnikov Ya A, Braginetz Yu.P, Vezhlev E O, Kuznetsov I A, Lasitsa M V, Seminikhin S Yu and Fedorov V V 2014 *Pis'ma v ZhETF* **100** 555
[17] Frank A I, Geltenbort P, Jentschel M, Kulin G V, Kustov D V and Strepetov A N. 2012 J*ournal of Physics: Conference Series* **340** 012042.
[18] Frank A I , Naumov V A 2013 *Phys. At.omic Nucl.* **76** 1423
[19] Landau L D and Lifshitz E M *Quantum mechanics. Non-Relativistic Theory,* 3rd edition, revised and enlarged (Pergamon Press, 1991), section 17.
[20] Landau L D and Lifshitz E M *Electrodynamics of continuous media,* 2nd edition revised and enlarged by Lifshitz E M and Pitaevskii L P (Pergamon Press, 1984), section 85.
[21] Yeh J 1965 *J. Appl. Phys.* **36** 3513
[22] Frank A I 2014 *JETP Letters* **100** 613
[23] Summhammer J., Badurek G, Rauch H, Kischko U and Zeilinger A, 1983 *Phys.Rev. A* **27**, 2523
[24] Badurek G, Rauch H, and Summhammer J, 1988 *Physica B* **151**, 82
[25] Golub R and Gähler R, 1987 *Phys. Lett.* A **123**, 43
[26] Gähler R, Golub R, and Keller T, Physica B 1992 **180-181**, 899
[27] Besenböck W, Göhler R, Hank P, *et al.,* 1998 *J. Neutron Res.* **7**, 65
[28] Kozlov A V and Frank A I, 2005 *Phys. At. Nucl.* **68** 1104
[29] Frank A I and Kozlov A V, 2009 *Physica B* **404**, 2550
[30] Baryshevskii V G, Cherepitsa S V, and Frank A I, 1991 *Phys. Lett. A* **153**, 299
[31] Baryshevskii V G and Podgoretsky M I, 1965 *Sov. Phys. JETP* **20**, 704 704
[32] Abragam A, Bachella G E, Glattli G L, et al., 1973 *Phys. Rev. Lett.* **31**, 776
[33] Frank A I, Bondarenko I V, Kozlov A V, *et al.,* 2001 *Physica B* **297**, 307
[34] Mezei F (editor) 1980 *Neutron Spin Echo*, Springer-Verlag, Heidelberg

Institute of Physics
Dirac House, Temple Back
Bristol BS1 6BE UK

ISSN: 1742-6588
ISBN 978-1-5108-2422-5

CURRAN ASSOCIATES INC.
proceedings.com

International Seminar on Interdisciplinary Problems in Additive Technologies 2015

IOP Conference Series: Materials Science and Engineering
Volume 140

Tomsk, Russia
16 - 17 December 2015

IOP
Institute of Physics